About the author

Anna Stavrianakis is a lecturer in international relations at the University of Sussex. Her main research interests are NGOs and global civil society; the arms trade and military globalisation; and critical approaches to the study of international security.

Taking Aim at the Arms Trade

NGOs, Global Civil Society
and the World Military Order

Anna Stavrianakis

Zed Books
London & New York

Taking Aim at the Arms Trade: NGOs, Global Civil Society and the World Military Order was first published in 2010 by Zed Books Ltd, 7 Cynthia Street, London N1 9JF, UK and Room 400, 175 Fifth Avenue, New York, NY 10010, USA

www.zedbooks.co.uk

Designed and typeset in Bembo by Kate Kirkwood

Index by Rohan Bolton, Rohan.Indexing@gmail.com

Cover designed by David Bradshaw

A catalogue record for this book is available from the British Library
Library of Congress Cataloging in Publication Data available

ISBN 978 1 84813 268 9 hb
ISBN 978 1 84813 269 6 pb
ISBN 978 1 84813 270 2 eb

For Brownie

Contents

Acknowledgements

This book is the product of engagement with a variety of people over several years. The strengths of it, such as they are, would not have been possible without those people, although I alone am responsible for its weaknesses. I am grateful to the many NGO staff members and volunteers, and civil servants involved in arms export licensing, who have generously discussed a variety of arms trade and control issues with me over time and, in some cases, commented on the manuscript. At the University of Bristol, where this academic project began life, Rob Dover and Ruth Blakeley provided moral and intellectual support, and I am grateful to Jutta Weldes and Richard Little for their guidance. More recently, at the University of Sussex, colleagues and friends have provided a congenial and stimulating environment in which to work. It is a pleasure and a privilege to be part of such a community. Spanning these geographical and temporal locations has been the invaluable combination of support and criticism from Tarak Barkawi, and stimulating dialogue with Neil Cooper, Keith Krause and David Mutimer.

I am very grateful to Tarak Barkawi, Ruth Blakeley, Rob Dover, Laleh Khalili, Jasna Lazarevic, Jan Selby, Martin Shaw and the anonymous reviewer for commenting on some of all of the manuscript. Some of the material in this book was presented at various BISA and ISA annual conferences, and at the LSE's Centre for Civil Society and Centre for the Study of Global Governance; thanks to the participants at these events for their constructive feedback. For their guidance during the publication process, thanks are due to Ellen Hallsworth and Ken Barlow at Zed Books. The original project on which this book is based was made possible by an ESRC PhD studentship and a Postdoctoral Fellowship.

I am indebted to my family and friends for their support over the years, and for their restraint in asking when the book – affectionately known as *Chicks and Guns* – will be finished. I hope that what follows will provide some explanation of what I've been doing with my time. This book is dedicated to our dear friend Joanna Brown, who died suddenly in August 2008. May her generosity, adventurousness, determination, kindness and humour continue to inspire the rest of us.

Abbreviations

ADS	AeroSpace, Defence and Security Industries (UK)
AeIGT	Innovation and Growth Team
BASIC	British American Security Information Council
BERR	Department for Business, Enterprise and Regulatory Reform
BIS	Department for Business, Innovation and Skills
CAAT	Campaign Against Arms Trade
CHAD	Conflict and Humanitarian Affairs Department
CHASE	Conflict, Humanitarian and Security Department
CND	Campaign for Nuclear Disarmament
DDR	disarmament, demobilisation and reintegration
DESO	Defence Export Services Organisation
DfID	Department for International Development
DSEi	Defence Systems and Equipment International
DSO	Defence and Security Organisation
DTI	Department for Trade and Industry
ECGD	Export Credits Guarantee Department
EDA	European Defence Agency
ENAAT	European Network Against Arms Trade
EU	European Union
FCO	Foreign and Commonwealth Office (UK)
GDP	gross domestic product
HDI	Human Development Index
IANSA	International Action Network on Small Arms
ICAO	International Civil Aviation Organization
ICBL	International Campaign to Ban Landmines
IGO	inter-governmental organisations
IHL	international humanitarian law
ISIS	International Security Industrial Strategy

ITAR	International Traffic in Arms Regulations
JRCT	Joseph Rowntree Charitable Trust
LDC	Least Developed Country
MoD	Ministry of Defence (UK)
MP	Member of Parliament
NATO	North Atlantic Treaty Organisation
NDIC	National Defence Industries Council
NGO	non-governmental organisation
NISAT	Norwegian Initiative on Small Arms Transfers
NOD	Non-Offensive Defence
NRA	National Rifle Association (US)
ODA	overseas development aid
OECD	Organisation for Economic Co-operation and Development
PRIO	International Peace Research Institute, Oslo
RAF	Royal Air Force
SFO	Serious Fraud Office (UK)
SIPRI	Stockholm International Peace Research Institute
SSR	security sector reform
UKTI	UK Trade and Investment
UKWG	UK Working Group on Arms
UN	United Nations
WFSA	World Forum on the Future of Sport Shooting Activities
WMD	weapons of mass destruction
WOOC	Work On Own Country

Introduction

On 6 December 2006, the United Nations General Assembly took a historic step in voting in favour of a proposal to pursue an international, legally binding Arms Trade Treaty. Recognising that the 'absence of common international standards on the import, export and transfer of conventional arms undermin[es] peace, reconciliation, safety, security, stability and sustainable development', Resolution 61/89 mandated the UN to establish agreed international rules for arms transfers (United Nations General Assembly 2006). Non-governmental organisations (NGOs), including Amnesty International and Oxfam, themselves heavily involved in promoting the idea and detail of an arms trade treaty, celebrated the vote as 'a strong indication that the global political will now exists to address the irresponsible and poorly regulated trade in arms, a trade which fuels conflict, results in gross human rights abuses and serious violations of international humanitarian law (IHL), destabilises countries and regions and undermines sustainable development' (Amnesty International 2007; also Control Arms 2006a).

Eight days later, the UK Serious Fraud Office suspended its investigation into allegations of bribery and false accounting by BAE Systems in relation to arms sales to Saudi Arabia. The UK government argued that continuing the investigation might have led to a withdrawal of Saudi diplomatic cooperation on security and intelligence issues and thus represented too great a risk to national security. News reports claimed that the Saudi government had threatened to suspend diplomatic ties with the UK and cancel a further proposed order for 72 Eurofighter Typhoon aircraft if the investigation was not halted (e.g. Leigh and Evans 2006, Percival 2006). In response, two NGOs, Campaign Against Arms Trade (CAAT) and The Corner House, launched a judicial review process against the government, arguing that the decision was unlawful on the basis of the OECD Anti-Bribery Convention, to which the UK is a signatory.

Since 2006, the UN process on the Arms Trade Treaty has led to an unprecedented number of states contributing to the UN Secretary General's

1

consultation process, broadly agreeing that human rights and international humanitarian law should be taken into account in international arms transfers. The US was the only government to vote against Resolution 61/89 in 2006; by late 2009, with a change of government, it had dropped its opposition and announced that it was 'committed to actively pursuing a strong and robust treaty that contains the highest possible, legally binding standards for the international transfer of conventional weapons' (Clinton 2009). In October 2009, 153 states agreed a timetable to establish a 'strong and robust' Arms Trade Treaty; only Zimbabwe voted against the proposal, with 19 states abstaining (Control Arms 2009). NGOs have played a pivotal role in garnering state support and pushing for the strongest possible wording of any eventual treaty.

In the UK meanwhile, the case mounted by CAAT and The Corner House was interrupted by the discovery that BAE Systems, the main company involved in the UK–Saudi arms relationship, was in possession of privileged and confidential legal advice to CAAT from its solicitors. It emerged that BAE Systems had, over a number of years, paid private intelligence companies to infiltrate and collect information on the campaign group. When the judicial review resumed, the NGOs initially won their case; however, the Law Lords overturned their victory on appeal. The Serious Fraud Office investigation remains discontinued in the UK, although the US Department of Justice has since begun its own probe.

These two episodes, coming so close to one another, paint starkly different portraits of the arms trade and, by extension, international security. In the first, the problems posed by the arms trade revolve around the need to promote human rights and international humanitarian law, prevent terrorism, and recognise the links between conflict, security and development. The problem is the irresponsible, that is, poorly regulated, arms trade. Many governments, encouraged by NGOs, have signed up to an explicitly international attempt at control, one that acknowledges the internationalisation of arms production and the interconnections between Northern and Southern (in)security. In the second scenario, in contrast, the problem is the state-sanctioned trade in arms, in particular that between one of the world's largest arms-exporting states, the UK, and one of the world's largest arms-importing states, Saudi Arabia. The relationship that arms-producing companies enjoy with governments is a significant feature of this account. Fuelled by corruption and tainted by secrecy, the arms trade is deemed to undermine the rule of law as well as human rights, peace and security, with states and arms capital working in tandem to protect their interests. While the vote in favour of an Arms Trade Treaty was a vote for an international and legally binding initiative and has been actively promoted by the UK government, a week later the UK undermined some of those very principles.

As well as contrasting portraits of the problem posed by the arms trade and the role of governments in it, the two episodes paint different pictures of civil society activism in world politics. The NGOs pushing for an Arms Trade Treaty – led by Amnesty International, Oxfam and the International Action Network on Small Arms (IANSA) – worked as part of an international coalition comprising Northern and Southern NGOs, engaging closely with governments around the world (arms-producing as well as non-producing and non-supplying) in trying to promote the toughest possible treaty. CAAT and The Corner House took a different tack, challenging a government decision in a national court, opting for confrontation rather than persuasion.

Such contrasts generate the key questions that animate this book. How do NGOs understand and campaign on the arms trade, with what strategies and to what effect? Further, what are the implications of this for how we think conceptually about global civil society, of which NGOs are emblematic actors? The aim of this book is critically to assess NGO arguments and strategies in relation to the arms trade, inquiring as to the adequacy of their responses as interventions in international relations. It asks what practices of international relations are produced, reproduced and/or challenged through NGO activity on the arms trade.

The International Campaign to Ban Landmines (ICBL) was a pivotal moment in post-Cold War civil society activism on weapons issues, achieving a ban on the use, development, production, acquisition, stockpiling, retention and transfer of anti-personnel mines (see ICBL, no date). Key features of the process were the NGO campaign, the role of small and medium powers, and the fact that it took place outside the UN system; both the ICBL and its co-ordinator, Jody Williams, were awarded the Nobel Peace Prize in 1997. Williams argued that the landmines campaign process 'has added a new dimension to diplomacy and hope for its wider applicability', describing the new model of partnership between civil society and small and middle powers as 'a new kind of "superpower" in the post-Cold War world' (Williams 1999; Clegg 1999).

Since then, civil society activism on weapons issues has become increasingly visible, with NGOs playing a key role in the adoption of the UN Programme of Action on Small Arms and Light Weapons in 2001 and the Convention on Cluster Munitions in 2008. To complement this role in international policy development, they have documented and publicised the effects of the arms trade on human rights, development, conflict and international security. They have criticised the abuse of military and security equipment, and the international transfers that facilitate it, in cases as diverse as: the use of lethal force against protestors in Guinea and Myanmar; armed violence against women in Guatemala; extrajudicial killings and war crimes in Somalia and the Democratic Republic of Congo; war crimes by both Israel and Hamas; corruption in military procurement in South Africa, India and Saudi Arabia; the cost of conflict in Liberia and Burundi; and poor export licensing decisions by the UK government in relation to countries in conflict, suffering human rights violations and/or underdevelopment, including but not limited to Indonesia, Nepal, Tanzania and Turkey (for a non-exhaustive survey, see Amnesty International 2009a, 2009b; CAAT 2006a; Oxfam 2008a; Saferworld 2007a). The list of abuses of military and security equipment around the world is, literally, endless.

Scholars are increasingly analysing the pivotal role played by civil society actors in post-Cold War weapons policy and practice (e.g. Anders 2005; Brem and Rutherford 2001; Cameron et al. 1998; Garcia 2006; Hubert 2000; Krause 2002; Price 1998). Such accounts draw our attention to the agency of actors beyond the state, arguing that civil society can exercise considerable influence in setting agendas, socialising other actors, generating and promoting norms, and influencing discursive and policy change, even on 'hard' security issues such as weapons policy. They also broaden our conception of security by drawing attention to the variety of ways in which weapons proliferation can be constructed as a threat: to human security, to human rights and to development as well as to the state. This book focuses on the agency of a community of NGOs, situating them as key actors in a purportedly globalising civil society. It does not focus on a particular category of weapons such as landmines, cluster munitions or small arms, but rather on the wider range of conventional weaponry, asking how the various NGOs understand the arms trade to be a problem in world politics and assessing the effects of their interventions.

The cast of characters

This book focuses on the main UK-based NGOs active on arms trade issues since the end of the Cold War: Amnesty International (both the UK section and International Secretariat); British American Security

Information Council (BASIC); Campaign Against Arms Trade (CAAT); the International Action Network on Small Arms (IANSA); International Alert; Oxfam (both Oxfam International and the Oxfam GB section); and Saferworld. These are not the only non-governmental actors working on arms trade issues, either within the UK or internationally. Other actors include: NGOs such as Human Rights Watch (New York), the Centre for Humanitarian Dialogue (Geneva), War on Want (London) and SaferAfrica (Pretoria); peace organisations including the UK-based Campaign for Nuclear Disarmament (CND), Peace Pledge Union and Fellowship of Reconciliation (although the latter also has an international presence); and research centres such as the Geneva-based Small Arms Survey, the Stockholm International Peace Research Institute (SIPRI), the Federation of American Scientists, headquartered in Washington, DC, and the Oslo-based International Peace Research Institute (PRIO). UK-based actors are significant players within this wider community: the IANSA secretariat, which coordinates 800 organisations working on gun violence in 120 countries, is based in London, as is the International Secretariat of Amnesty International, while Oxfam's is based in Oxford. The UK sections of Amnesty International and Oxfam, along with Saferworld, are key contributors of research and strategic expertise on both national and international arms control matters.

The ongoing importance of nation states to the arms trade, and the NGOs' own experience of campaigning on national control issues, are both reasons to retain a focus on a nationally based group of actors, while not ignoring their international links. The arms industry is internationalising, but its main customers are state governments, which also regulate their activity. The arms trade has an ambivalent relation to wider processes of trade liberalisation, and is heavily marked by national protectionism. The NGOs themselves have concentrated on variously encouraging or challenging the UK state to improve its policy and practice. Since 2003 however, Amnesty, IANSA, Oxfam and Saferworld have broadened their focus to try to improve international controls via the Arms Trade Treaty. And even 'international' NGOs, such as Amnesty and Oxfam, are still structured in terms of nationally based sections.

CAAT is the only one of the seven organisations to focus specifically and solely on the conventional arms trade, and is the most focused on the role of the UK. This is a legacy of its formation in 1974 as a campaign network of peace organisations concerned about the growth of the arms trade after the 1973 Middle East War, including CND, Peace Pledge Union, Fellowship of Reconciliation and Quaker Peace and Social Witness. While CAAT has since become an organisation in its own right, its original creators continue to send representatives to its Steering Committee, giving input from the wider peace movement to CAAT's campaigning.

Saferworld also grew out of the peace movement, specifically the Nuclear Freeze campaign. After the Intermediate-Range Nuclear Forces Treaty of 1987 eliminated this category of weapons, Freeze activists moved into a wider post-Cold War security remit. There was a concomitant shift of strategy from campaigning to policy advocacy, which marks Saferworld out from CAAT, despite their shared Cold War peace movement background. Saferworld's work on conventional export controls was prompted by the first Gulf War and played a significant role in the organisation's expansion. In the mid 1990s there was another shift, towards conflict prevention, at which point Saferworld started to focus on small arms and also started to approach the UK government for funding. In the late 1990s Saferworld's first in-country programmes on small arms control were established. BASIC was also established in 1987, with a focus on the full spectrum of military production and trade, from nuclear and other non-conventional weapons to conventional export controls and small arms. This makes it the only NGO to work across the full spectrum of weaponry; it is also unique in its transatlantic focus.

Amnesty International and Oxfam have tackled arms trade issues as they relate to their respective remits of human rights and poverty since the mid 1990s. For Amnesty, the Rwandan genocide was a key example of the role of the arms trade in crimes against humanity that prompted its activism on the issue (Amnesty International 1995). Between 2000 and 2002, the two NGOs ran a joint campaign called 'Aim Higher for Tough Arms Controls', calling for stronger legislation and a robust system of end-use monitoring on UK arms exports. By 2002 the need to 'Curb the flow of arms that fuel conflict' featured in Oxfam's ten-point plan for international action (Oxfam 2002a). International Alert, a peacebuilding organisation, was established in 1986 by human rights advocates led by the former Secretary General of Amnesty International, Martin Ennals, 'in response to growing concerns expressed by those working in international development agencies, human rights organisations and those involved in the issues of ethnic conflict and genocide' (International Alert, no date). And IANSA, which calls itself 'the global movement against gun violence', was established in 1998 by a mix of Northern and Southern organisations, many of whom had been active in the ICBL.

There are significant links between the organisations, with a division of labour and subject expertise, sharing of information and a significant degree of personnel exchange over the years. At various times, all of the NGOs bar CAAT have been involved in the UK Working Group on Arms, which seeks to pool NGO resources and expertise to improve UK government policy and practice, and/or the Control Arms Campaign, which began in 2003. This campaign is led by Amnesty, Oxfam and IANSA and calls for an Arms Trade Treaty that would codify states' existing responsibilities into an international legally binding treaty. The campaign also has an International Steering Committee that aims for balance in both geography and specialism; it comprises eleven members in addition to the three lead organisations, including Saferworld, organisations from Kenya, Costa Rica, Brazil, Trinidad and Tobago, Canada and the US. However, the majority of campaign reports are written by the staff of UK-based NGOs, in particular Amnesty, Oxfam and Saferworld.

CAAT's links are with a slightly different set of organisations; as well as the peace organisations that sit on its Steering Committee, it works with actors as diverse as War on Want, Tapol (the Indonesia human rights campaign) and Speak (a Christian student prayer and campaign group). It is also a member of the European Network Against Arms Trade (ENAAT), comprising nationally based groups and individuals from across 13 European countries working against the threat to international peace and security posed by the arms trade. In general, CAAT and its allies tend to have a more confrontational strategy and argument about the arms trade and related issues than the UK Working Group/Control Arms coalition. But all seven of the NGOs are linked, internationally and thematically, to other organisations, making them a key subset of a wider community working on arms issues.

There are thus shifting relationships and coalitions within the NGO world – both in the UK and internationally – and, of course, the counterpart to this of tensions and power struggles. NGO objectives, understandings of the arms trade, strategies and impacts vary; the aim of this book is to understand their activity, individually and as a whole, and to explore the practices of international relations that are reproduced and/or challenged through their activism. There are a plethora of definitions and usages of the term 'NGO' and of their perceived function in world politics. Distinctions can be made according to their size, income, function (e.g. relief, advocacy, campaigning), scope (local, national, international, transnational, etc.), audience (grassroots, elite), membership, across a range of issues (human rights, social justice, welfare, the environment), objectives (from conformists to transformists) and strategy (from insider to outsider approaches). Labels such as grassroots organisations, community-based organisations, civil society organisations, citizen groups and voluntary sector organisations abound in the literature as equivalent terms. What unites many or most of these conceptions is that NGOs are: independent of government, although they may receive government funding; usually reliant on voluntary contributions for a significant proportion of their income; and voluntary, although many NGOs employ at least some permanent staff. It is in this sense – given that all seven organisations are formally independent of government, not-for-profit, and organised (they have offices and budgets, and they employ permanent staff) – that they are included here under the label of 'NGOs'. Exploring the differences, complementarities and tensions between their objectives, strategies and effects is the task for the remainder of this book.

The argument

NGOs have been important players in post-Cold War conventional arms control. They have been highly critical of government policy and practice, and have exercised agenda-setting power in encouraging greater attention to the role of human rights, development and conflict prevention concerns in arms export licensing. In particular, NGO activism played a key role in: the agreement of the European Union Code of Conduct on Arms Exports in 1998 and its shift to a legally binding EU Common Position in 2008; the wording of the UK's 2002 Export Control Act, the first legislative change in this issue area since 1939; promoting increases in transparency brought about through regularised reporting by EU member states; pushing for controls on activities such as brokering and transportation that are pivotal to an increasingly globalised arms trade; shutting the UK Defence Export Services Organisation (DESO) in 2007; and promoting an Arms Trade Treaty. As such, they seem to be emblematic of a global civil society that promotes progressive, non-violent change in world politics. However, there are a number of reasons to be sceptical of such an optimistic judgement.

As is argued in Chapter One, how we conceptualise global civil society affects our perspective on its prospects of success. Global civil society is conventionally understood as a sphere separate from both the state and the market, one that is increasingly global, progressive and non-violent. This is a distinctively liberal understanding and one that ultimately hampers our understanding of the possibilities contained in global civil society.

Drawing on postcolonial and Marxist critiques of global civil society, and the Gramscian concepts of hegemony (widespread consent for the general direction of social life, backed up by state-sanctioned coercion) and counter-hegemony (a rupture in the naturalisation of hegemony, and the sowing of the seeds of an alternative order) (see Gramsci 1971, 12; also Cox 1983; Hall 1986), the chapter explores the ambiguities, tensions and trade-offs associated with NGO activity. Civil society is central both to the maintenance of and potential challenge to a hegemonic status quo, in which dominant ideas gain consent and become naturalised, setting the parameters of understanding and action. The production and reproduction of social power and relations of domination are always contestable (Rupert 2005, 209) and the analysis of NGO activity in the chapters that follow demonstrates the extent to which NGOs play a role in perpetuating or challenging dominant narratives and practices of the arms trade. Conceptual critique is important to our understanding of NGO interventions in international relations as it provides us with a means of interpreting the effects of NGOs' activism beyond success or failure defined in their own terms.

Chapter Two demonstrates the different ways that the arms trade is conceptualised as a problem by the various NGOs, distinguishing between reformist and transformist approaches. A reformist approach, epitomised by Amnesty, Oxfam and Saferworld, focuses on the impact of the arms trade where it is most keenly felt, in regions of the South where human rights violations, conflict and underdevelopment are rife. The reformists seek to mitigate the worst excesses of the arms trade through tighter governmental control and regulation. BASIC and CAAT agree that the effects of the arms trade are worst in the South, but have a different argument as to the nature of the problem and a different set of proposed solutions. For BASIC, while human security is threatened by small arms proliferation (and it worked on small arms and export control issues for many years), nuclear proliferation poses a greater existential threat to the planet more generally, and since 2007 the organisation has returned its focus to global nuclear disarmament. CAAT focuses on the impetus to the ills caused by the arms trade, which in its view is the close relationship between arms capital and the state. BASIC and CAAT can be understood as transformists as their analyses point to a radical overhaul of the arms trade as a system, as opposed to regulating away its worst excesses within the existing system. These different NGO understandings should be contextualised in terms of a world military order that remains inflected by North–South relations despite the end of the Cold War and the growing internationalisation of military production and trade.

That is, international arms exports are just one element of a wider system of military production and trade, and this system remains dominated by the US and, to a lesser but still significant extent, Russia and west European states.

Chapter Three analyses the strategies and campaign histories of the seven NGOs, outlining the spectrum of insider and outsider strategies in play. It contextualises NGO activity against the capitalist structuring of civil society and the integration of arms capital into the UK state. Not only do we need to understand the cumulative effect of the NGO community as a whole, we also need to understand the differential integration of even the most insider NGOs compared to arms capital. This helps us to map the terrain of actors on arms trade issues and to assess the potential for change more effectively.

Chapters Four, Five and Six examine NGO activity in relation to three of the main trends in play in the post–Cold War era, namely intra-Northern production and trade; North–South transfers; and small arms proliferation. Chapter Four focuses on NGO activism in relation to economic arguments, domestic procurement and UK military posture, as well as transatlantic and European defence collaboration, as key examples of intra-Northern military production and trade. NGO research, advocacy and campaigning on economic arguments around arms exports in the UK has often been transgressive and has, over time, led to a degree of policy change. But more generally, these issues are notable both for the lack of attention they receive from NGOs and for the way in which they are largely constructed as a defence-industrial rather than a security issue. Even when security concerns are incorporated, these focus on the risks of exports to the South rather than processes of Northern militarisation. The partial exceptions to this are BASIC and CAAT, whose respective remits of transatlantic security and national involvement in the arms trade facilitate their interventions on these issues, at times in a transgressive fashion.

Chapter Five examines NGO activity in relation to arms transfers to the South, focusing on the risks to human rights and development. These issues are a key concern for Amnesty International, Oxfam, Saferworld and CAAT. Despite the differences in argument and strategy adopted by the various NGOs, a common effect of their activity is to represent the problems of the arms trade as internal to the South, in that they are deemed to result from misallocation of resources, lack of training and so on. The same can be said of NGO activism in relation to conflict and small arms, analysed in Chapter Six. Amnesty International, IANSA, International Alert, Oxfam and Saferworld have been particularly active on small arms issues, which are emblematic of the liberal conflict–security–development framing that has come to dominate post-Cold War international security theory and practice and serves to reproduce the South as a site of intervention in international politics. NGOs' focus on the effects of the trade on the South, in particular the egregious results of small arms proliferation, stems from a progressive impulse to mitigate against harm and they are critical of the role of suppliers as well as recipients. However, this normative impulse, and NGOs' interventions that stem from it, serve to disconnect organised violence in the South from the wider world military order, a spectrum of militarisation that is marked by North–South hierarchy, and the historical and contemporary processes of capitalist globalisation of which violent social relations are often a part.

Overall then, the optimistic assessment of NGO activity needs to be qualified. NGO activism sidelines intra-Northern military production and trade as part of the problem of the arms trade; and the focus on transfers to the South that damage human rights and development, and the growing international control regime in relation to small arms, are both based on liberal assumptions that serve to disconnect organised violence in the South from its historical and international context, and from wider processes of organised violence through the world military order. The focus on the South at the expense of either the North or North–South relations serves to construct the problem of the arms trade as internal to the South and, crucially, reproduces the South as a site of intervention. NGOs thus participate in and naturalise a hierarchical world military order despite their (self-)image as progressive actors. As is argued in Chapter Seven, this begs a rethinking of our understanding of the power of NGOs as global civil society actors and their transformatory potential. While based on seemingly universal goals of the promotion of human rights, development and conflict prevention, the partial nature of their activity in relation to the arms trade serves to reproduce a profoundly hierarchical and violent world military order.

At the broadest level, both the arms trade and NGO activity in relation to it are understood in this book as a means of interconnection and mutual constitution in international relations, shaping both the North and the South (Barkawi and Laffey 2002; Barkawi 2006). For example, in the judicial review case mentioned above, efforts by two small, resource-poor campaign organisations to challenge the relationships between the UK and Saudi governments and between BAE Systems and the UK government led to two High Court judges ruling that they were intervening to uphold the rule of law against threats made by representatives of another state. The UK government has meanwhile embarked on a programme of constitutional renewal, enshrining in the powers of the Attorney General the ability to halt investigations on grounds of national security and prevent such challenges being brought again. Thus, not only has the UK's arms relationship with Saudi Arabia contributed to a particular path of military development in the latter country, it has also prompted constitutional change at home, striking at the heart of the British legal and political system.

Through its support for an international Arms Trade Treaty, the UK – consistently one of the world's top five arms-exporting states – represents itself as intervening on behalf of those millions of people who live in daily fear of armed violence. Focused in practice on small arms predominantly, this effort is both a form of intervention in the South through small arms control and a means of creating a benevolent self-identity at home. The arms trade, and NGO activity in relation to it, are thus a means of interconnection in international relations, a mechanism through which relationships are produced, reproduced and, sometimes, challenged.

This emphasis on social relations means that categories such as 'North' and 'South' do not have a prior existence as unproblematic entities; rather, they are produced and reproduced through discourse and practice (Doty 1996, 1). That is, the North and South are 'imagined geographical space[s]' (Duffield 2002a, 1052) rather than objective labels to be attached to particular places (also Ó Tuathail 1996). They are most often constructed in a relationship of domination and subordination, in which the South is understood as dangerous, barbarian, weak, or somehow otherwise lacking, in contrast to the rational, strong, benevolent North (Doty 1996; Escobar 1995; Said 1978). In the context of the arms trade, the dominance of the US, European states and also Russia within the contemporary world military order is what makes the category 'North' more appropriate than 'West'. The South, meanwhile, comprises the non-European world, or what is widely called the developing or Third World. The end of the bipolar Cold War system largely made the labels of 'First', 'Second' and 'Third' worlds redundant, but has also led to a repopulation of various categories, such that, for example, the Balkans are constructed as Southern in arms trade and control debates.

This argument is made from a position of critical sympathy to the NGO cause. Military production and trade are undoubtedly out of control. Their effects are widespread and devastating, contributing to ongoing militarised, imperialist, racialised and gendered practices of coercion, violence and intimidation in an internationalising capitalist state system. The critique being made in this book is aimed not at individuals, but is an analysis of the disciplining and structuring of civil society that prevents more transgressive action against the arms trade that challenges the scale of, and asymmetries in, global military production as well as controversial exports. There are several instances in which I critique NGO arguments in terms of the power relations they reproduce, even if for reasons of strategy it makes sense for NGOs to make them. The aim is to highlight the distinction between intention and effect, to ask what power relations are being reproduced even by the most well-intentioned of actors.

This is the space that academia provides, allowing scholars to ask questions that cannot easily be addressed in the NGO world. The community of NGOs working on arms issues is a small one, and its participants are mostly known to each other; many of them have worked for more than one of the NGOs under analysis here. Indeed, over the past seven years I have at various times worked voluntarily with more than one of the NGOs under analysis, starting at Saferworld in 2002 and ending at CAAT in 2009. Fieldwork for this book was conducted over the past seven years, in the form of interviews with over fifty current and former NGO staffers and volunteers and UK civil servants involved in arms export licensing and promotion, participant observation at a variety of practitioner events, and documentary analysis. Staffers' institutional memory of individual organisations and the community as a whole has been invaluable. To emphasise that the critique is not aimed at individuals, I have attributed interview material by the speaker's organisation and/or role only, unless the sensitivity is such that complete anonymity is more appropriate. I use real names only when individuals are on record as part of a public commentary or intervention.

1 • Conceptualising Global Civil Society

The concept of global civil society has borne the hopes of the post-Cold War world on its shoulders. It is widely deemed to be a pacific site of emancipatory potential, if not practice: our best hope for the promotion of human rights, development and good governance. While most often associated with 'soft' security issues such as environmental degradation, poverty, women's and indigenous rights, or health issues such as HIV/AIDS, global civil society also has a track record of tackling 'hard' military or weapons issues. Cold War peace movement campaigns to ban nuclear weapons have outlived the end of superpower rivalry, and new campaigns against landmines, cluster munitions, small arms, depleted uranium and the use of child soldiers have been launched since the 1990s. Commenting on the role of NGOs in the Ottawa process that banned landmines, Canadian Foreign Minister Lloyd Axworthy argued that 'one can no longer relegate NGOs to simple advisory roles...They are now part of the way decisions have to be made' (in Brem and Rutherford 2001, 175). Increased political, popular and scholarly attention to civil society activism is predominantly based on a liberal account of the post-Cold War and now post-9/11 context of globalisation and security, often understood as the intensification of interaction between states and communities in an age of global tele–communications, the rising salience of non-state actors, non-traditional security threats, the increasing spread of norms, and the emergence of global governance. However, this chapter seeks to problematise this optimistic assessment of NGO and civil society activism in four main ways.

First, the common definition of global civil society as a non-state, non-market sphere fails adequately to situate it in the historical and social setting of a global capitalist state system. In privileging the agency of global civil society actors such as NGOs and untethering them from the structures that ground them, the constraining and enabling factors that necessarily affect their prospects of success are sidelined. Second, NGOs often claim to be, and

14

are widely understood to be, driven by progressive or emancipatory values. However, the disciplining of NGOs' activity through their structural position within civil society is such that this normative judgement should be more ambivalent than is usually recognised. Third, accounts of global civil society often rest on a problematic conception of the global that hides not only the uneven geographies of civil society but also the imperial (re)ordering of international relations, and serves to depoliticise the transformatory and universalising urge inherent within liberalism. Fourth, the emphasis on the non-violent nature of global civil society sidelines the violence of capitalism and the state system and disciplines dissent within the NGO and activist world.

These four limitations of the mainstream literature on global civil society stem from a Eurocentric liberal orientation to international relations that fails to recognise civil society as a distinctly modern phenomenon associated with the rise of European capitalism, or, if it does recognise this, continues to apply a historically and geographically specific concept universally. This results in the naturalisation of an idealised conception of state–society relations based on the European experience. It also raises the question of how liberal scholars understand global civil society to operate in non-European and postcolonial contexts, and how NGOs understand global civil society to operate in the regions and countries in which they work. A key effect of both the global civil society literature and NGO practice is to shore up relations of hierarchy between North and South as they facilitate intervention in the South by a network of state and non-state actors predominantly from the North. The conceptual critique set out here lays the ground for the empirical analysis of NGO activity that follows in the rest of the book. Despite their (self-)image, the activity of NGOs on the arms trade is often illustrative of these trends.

Global civil society as a non-state, non-market sphere

Most contemporary accounts of global civil society define their object of study as a realm distinct from both the state and the market. It is referred to variously as a '"third system" of agents, namely, privately organised citizens as distinguished from government or profit-seeking actors' (Price 2003, 580), a 'third force' that 'includes only groups that are *not* governments or profit-seeking enterprises' (Florini and Simmons 2000, 7, italics in original), or the organised expression of the public sphere, which resides between state and society (Castells 2008, 78-9). It is thus located 'between' the economy and the state (e.g. Anheier *et al.* 2001, 17; Cohen and Arato 1992, ix).

Following on from this definition, the main focus is on actors such as non-profit groups, charities, social forums and movements, and informal associations; a key role is played by NGOs (Castells 2008, 83-6; Florini and Simmons 2000, 13; Kaldor 2003a, 13; Lipschutz 1992, 390; Scholte 2004, 214-15; Shaw 1994, 650). While there is considerable internal diversity within a broadly liberal approach (from the cosmopolitanism of Mary Kaldor and associated cosmopolitan democracy theorists, to the global governance approach of Jan Aart Scholte, to the constructivism of Richard Price and others working on issues around norms in international relations), there is a dominant 'distinctive liberal theoretical statement' that 'privileges the role of agency, namely transnational civil society activists', thus challenging other theoretical approaches that privilege other agents, or structures (Price 2003, 601).

In the standard historical tracing of the concept of civil society, scholars often cite the differentiation of civil society from the state as one of the key markers in the development of the concept (e.g. Anheier *et al.* 2001, 13; Shaw 1994, 647). As Lipschutz puts it, whilst Locke and Marx share the conventional definition of civil society as existing 'in some twilight zone between state and markets, engaging in activities that constitute and reproduce the fabric of everyday social life', they deploy it 'to differing conclusions' (2005, 758). However, while this difference is initially recognised, there is a slippage in subsequent usage of the concept of civil society in liberal accounts, with the separation between the state, market and civil society being naturalised. That is, scholars often forget that the differentiation of civil society from the state and market is purely methodological, so as to allow 'a serious thematisation of the generation of consent through cultural and social hegemony as an independent and, at times, decisive variable in the reproduction of the existing system' (Cohen and Arato 1992, 143; see also Robinson 1996, 352-4). There is a tendency in the liberal literature for this methodological separation to be taken as a substantive separation, with the effect that the structural constraints on actors such as NGOs are occluded.

In contrast, Marxist accounts of civil society emphasise that the modern separation of public from private, state from market, is purely formal. This means that civil society contains the market and is riven by class inequalities (e.g. Colas 2002). In order to participate in the public realm of the state as citizens, people have to abstract from their real lived selves. The state, or political life, in which differences between individuals are seen merely as social differences with no political significance, rests on the inequalities of civil society (Marx 1975, 211-41). Contemporary Marxist scholars such as Ellen Meiksins Wood argue that any adequate account of civil society must expose 'the relations of exploitation and domination which irreducibly *constitute* civil society' because Gramsci's concept of civil society 'was unambiguously intended as a weapon against capitalism, not an accommodation to it' (Wood 1990, 63, 74, emphasis in original; also Colas 2002, ch. 2). Marxist accounts of NGO activity often emphasise its role in undermining emancipatory struggle in the South and reproducing imperial relationships through a failure to tackle structural causes of poverty (e.g. Hearn 2001, 2007; Manji and O'Coill 2002; Wallace 2003).

While there is sometimes a tendency to overdetermine this feature of civil society in an instrumentalist or functionalist manner (e.g. Petras 1999), the key lesson is neatly articulated by Colas, who argues that the predominantly consensual approach taken by international NGOs 'is not simply a matter of political choice or preference on the part of civil society agents, but rather a structural property of the current relation between global civil society and global governance' (2002, 156). That is, there is a historically grounded structural bent to NGOs' political orientation, regardless of the intentions of those individuals working within them. This is part of the ambiguity of civil society: there is a disjuncture between NGO staffers' intentions and effects, between civil society's role in the propagation of hegemonic understandings and practices, and its potential role for resistance. While the structuring conditions of capitalist civil society shape and discipline the agency of actors such as NGOs, the precise dynamics in any given situation are a matter for empirical analysis.

The Marxist emphasis on the formal separation of the state from civil society and the historical specificity of its emergence can be productively read alongside postcolonial approaches. Civil society emerged as a geographically and temporally specific phenomenon in relation to the emergence of capitalism. Its application across space and time as a potentially universal emancipatory category is therefore problematic. As Chatterjee argues, the history of state–civil society relations 'is intricately tied to the history of capital', meaning that 'the concepts of the individual and the nation-state both become embedded in a new grand narrative: the narrative of capital' (1990, 123, 128). It is the 'moment of capital...global in its territorial reach and universal in its conceptual domain', that turns 'the provincial thought of Europe to universal philosophy, the parochial history of Europe to universal history' (Chatterjee 1990, 129).

Taking seriously the history lesson that the concept of civil society emerges with European modernity, in relation to a particular form of socio-economic and political community, means that the universal application of the concept of civil society is problematic. Thus, the common questions of why civil society is weak, fractured, absent or corrupted in the South assume that a particular historical experience is universally replicable; applying it to the 'global' level is an attempt to universalise a concept that is grounded in a specific historical context. This is not to deny that the concept has gained a foothold in the South and may be subject to reappropriation and reworking. Rather, the task of a postcolonial critique is to reveal the parochial nature of Europe, to disturb its self-image as (potentially) universal and to recall the centrality of capitalism to the emergence of civil society and its relations with the modern state.

Modern civil society is based on capitalist economic and social relations, which means that it exists 'not merely in opposition to the state but in relation to a certain form of state', namely one with 'effective rule based on representative institutions, supporting and supported by a system of rights' (Blaney and Pasha 1993, 6–7) in which the state holds a monopoly on legitimate violence. Analytically, the concept of civil society, like other concepts associated with modernity, 'is impossible to *think* of anywhere in the world without invoking certain categories and concepts, the genealogies of which go deep into the intellectual and even theological traditions of Europe' (Chakrabarty 2000, 4, emphasis in original). This raises the question of how the concept of civil society applies to the non-European world. Blaney and Pasha argue that there is a tendency for commentators to label 'informal economic activity' or 'any organised opposition to the state' as 'an emerging "civil society" and the bellwether of a democratic transition' (1993, 17). Similarly, Garland refers to the practice of 'looking for non-Western *analogues* to civil society' that appear to be 'the only viable option for a progressive politics'; this has the effect of naturalising the liberal origins of the concept (1999, 74, emphasis in original). She argues that NGOs 'appear to be almost natural institutional embodiments of the liberal conception' of global civil society, despite it being 'an ideologically charged ideal' (Garland 1999, 73). A postcolonial critique that highlights the particularity of the concept of civil society raises questions about scholars' and NGOs' understandings of the relations between state, market and civil society in the South.

A critical stance towards mainstream conceptions of civil society as a sphere substantively separate from the state and market opens up space for analysing

NGOs and the arms trade in a different light. Understanding the separation of civil society (which includes the market) from the state to be formal rather than substantive requires us to investigate empirically the relations between the state, arms capital and NGOs. As discussed in Chapter Three, rather than the three spheres (state, market, civil society) that would feature in a liberal account, it is more fruitful to think of two networks of actors, one comprising elements of the state and capital, the other consisting of other elements of the state allied to NGOs. This gives us a more nuanced appreciation of the social forces that NGOs can both draw on and need to challenge. It also helps us understand the consensual orientation of most of the NGO world as a structural property of the sector, rather than simply a strategic choice. While NGO staffers are often highly politicised and critical individuals, and as organisations NGOs exercise agency when designing and implementing campaigns, their choice of issue, target and mode of intervention is shaped by the capitalist social relations of which they are a part. Further, the critique made here opens up space for the argument made in chapters Five and Six, in that the different historical experience of state formation and capitalist globalisation in the South has led to social relations of coercion and resistance that do not match up to expectations based on the European experience. This encourages us to investigate the role of civil society activity in the South, and the relations between Northern and Southern actors.

Global civil society as the locus of progressive values

As agents of global civil society, NGOs claim to pursue progressive social change and are widely heralded as agents of a progressive politics, motivated by shared values such as altruism and a commitment to human rights, development and other common goods (e.g. Etzioni 2004; Florini and Simmons 2000, 7; Kaldor 2003a, 86; Lipschutz 1992). A caveat is often deployed to the effect that, when considered analytically, politically distasteful groups also count as global civil society, such that 'neo-Nazi hate groups…are just as much transnational civil society networks as are the human rights coalitions', for example (Florini and Simmons 2000, 231). However, most analyses of global civil society activism focus on its progressive variants. There is thus a broad normative impetus to this literature. This is problematic for at least two reasons: first, that civil society is structurally more ambiguous than we might like to think; and second, that ignoring 'uncivil' actors weakens our understanding of global civil society's progressive potential.

Mainstream accounts of global civil society tend to sideline the ambiguity of the concept and its role in buttressing the status quo as well as potentially supporting social transformation. Civil society has contradictions built into it: it can serve to naturalise and further entrench socially dominant forces but it can also (and perhaps simultaneously, depending on the concrete historical conditions) be the breeding ground for counter-hegemonic resistance and a new social order. In a Gramscian account of hegemony and counter-hegemony, the former is created and maintained through institutions such as the family, church, media and educational institutions (Cox 1983).

NGOs can be added to this list as they generate knowledge and contribute to the social construction of problems and appropriate responses to them, through their advocacy, lobbying and campaigning strategies. The key question is whether they can trigger a rupture in the naturalisation of hegemony and sow the seeds of an alternative order. Building counter-hegemony entails the creation of 'alternative institutions and alternative intellectual resources within existing society' and requires 'resisting the pressure and temptations to relapse into pursuit of incremental gains for subaltern groups within the framework of bourgeois hegemony' (Cox 1983, 165; see also Robinson 1996, 381). However, the disciplining effects of the capitalist structures of civil society make such resistance difficult for actors organised as NGOs. Indeed, as we shall come to see, some NGOs have a shared agenda with the state and, increasingly, the arms industry, with the effect that they pursue incremental change while accepting the parameters of the world military order.

This means that the claim that NGOs are agents of a progressive politics needs to be qualified somewhat, or at least investigated. Strategies against hegemony require transgression to be effective. That is, they must signal an 'assault on the way social norms, beliefs, inequalities and oppressions are reproduced' (Jordan 2002, 32). Effective strategies therefore make demands that 'cannot be met within existing structures': the changes that result from effective strategies do not leave society as it was before (Jordan 2002, 36). This is the distinction between incremental changes that leave the parameters of an issue untouched, and transgressive change that fundamentally alters the social landscape as well as generates concrete improvements.

Transgressive activity does not have to 'aim at all social institutions and structures simultaneously…different social institutions can be identified as a component of society that needs changing' (Jordan 2002, 37). But it is important to recognise the links between issues and create 'a chain of equivalences between all the democratic demands to produce the collective will of all those people struggling against subordination' (Mouffe 1998, 99). Crucially, transgressive change is always possible, as hegemony is never established once and for all: it requires ongoing political and cultural, as well as economic, practices to sustain it (Mouffe 1998, 91). The task is therefore to identify the weak spots in representations and practices, and exploit them in pursuit of changes that not only ameliorate the current situation but also change the terms of debate and understanding. The question of transgression finds empirical expression in the differences between reformist and transformist NGO approaches to the problem of the arms trade, discussed in Chapter Three, and the differences in the issues that NGOs focus on and the way they engage on them, as demonstrated in chapters Four, Five and Six.

It is important to avoid a monolithic understanding of hegemony. Given that trade in military equipment, militarism and the use of force predate modern capitalism, capitalist hegemony, whilst crucial to the analysis, cannot be the only hegemony in play. Other key hegemonic formations include militarism and states' use of force, given the centrality of war-making to state-making (Tilly 1985), as well as hierarchical North–South relations, a key feature of imperialism (Barkawi and Laffey 2002). So the question of whether NGOs challenge the various hegemonic social formations of capitalism, militarism and imperialism is key. How do they understand the arms trade, development, human rights and conflict prevention? How do these understandings relate to dominant accounts of state–capital relations, states' use of force, and North–South relations? And what strategies for change are facilitated by these understandings? These are some of the questions that animate this book.

Scholarly analyses also tend to relegate 'uncivil' actors to the margins, often using spatial metaphors to do so. Keane, for example, argues that global civil society 'contains pockets of *incivility* – geographic areas that coexist uneasily with "safe" and highly "civil" areas' (Keane 2003, 12, emphasis in original). We are told by Kaldor that cosmopolitanism is 'emerging side by side with the politics of particularism' (1999, 139). The task in responding to conflict is to find the 'islands of civility' that 'need to be taken seriously and given credibility by outside support', in order to generate 'an alliance between international organisations and local advocates of cosmopolitanism in order to reconstruct legitimacy' so that 'alternative forms of inclusive politics can emerge' (Kaldor 1999, 120-25; also 2003a, 5-6).

The aim of a safe, inclusive, civil and cosmopolitan social order sounds laudable, but the question remains of who defines the standards of acceptable behaviour. On this matter, liberalism has a chequered history, arrogating for itself the right to define and choose. Liberalism has historically been highly intolerant of illiberal or non-liberal social forms and practices (Jahn 2005; Mehta 1999; Parekh 1995). A key effect of the distinction between cosmopolitanism and particularism is to reproduce a binary between civil and uncivil, liberal and non-liberal and promote the former over the latter, while claiming to be neutral and universal.

Liberal scholars reject the idea that civil society is simply a vehicle for the imposition or promotion of Western liberal norms, however. They emphasise instead the 're-export' of languages and terms, meaning that the language of civil society is 'both *pluralised* and *globalised*' (Keane 2003, 38-9, emphasis in original). However, such accounts still locate progressive agency in cosmopolitan actors within global civil society – those who have internalised modern liberal subjectivity and personhood – and are blind to the imperial relationships fostered in the course of globalisation. Negotiation over concepts and meanings undoubtedly takes place but it does so in the context of an imperial hierarchy in world affairs, in which representational and physical encounters between the North and South are 'imperial encounters', that is, asymmetrical and also mutually constitutive (Doty 1996).

Imperial relations are still prevalent, not because of formal, territorial colonisation but because of continuing hierarchical representations of the North and South and the practices facilitated by them. We therefore need to ask what 'relations of power and struggle' are inscribed in the way people talk about global civil society (Mohanty 1997, 93). It is important to examine the discursive construction of global civil society and the ways in which NGOs (re)produce or challenge it. This raises the prospect of global civil society as a 'hegemonic project' in its own right (Pasha and Blaney 1998, 435). There is a transformatory, universalising urge within liberalism, one effect of which is that the concept of civil society gets 'unmoored from its very specific history in the West' and deployed in a 'normative manner', with the effect that 'the concept becomes a universal condition of possibility' (Bissell 1999, 124).

The emphasis on global civil society as a key locus of progressive politics is thus normatively rather than analytically driven. Not only is this intellectually problematic as it weakens the conceptual purchase of global civil society, it is also ultimately politically short-sighted even from within a liberal paradigm. It means that liberal scholars and practitioners fail to understand the practices they do not like, and thus devise inadequate responses. As

Pasha and Blaney argue, 'a failure to attend to the mutually constitutive relationship of civil society, capitalism, and the liberal state will misguide our assessments of the emancipatory possibilities of associational life' (1998, 420). Elsewhere, they argue that some efforts of indigenous peoples seeking to preserve isolation from the wider economy and political society, or efforts at theocratic constructions of political society, may be '*counter*-civil society movements' because they are not in line with the historical and theoretical specificity of the term 'civil society' (Blaney and Pasha 1993, 18, emphasis in original). Liberal approaches thus often ignore 'the vector of resistance politics'; that is, the marginalised may be actively resisting neoliberal globalisation rather than simply suffering its effects as victims (Mittelman 2005, 116).

A critique of the liberal assumptions behind the claim that global civil society contains the promise of a progressive politics qualifies the optimistic tone of much of the literature. While the empirical focus of this book is a set of pro-control NGOs that seem to embody such a progressive politics, there are three reasons to be sceptical. First, the structural ambiguity of civil society, discussed previously, requires us to consider arms capital and its associated lobby groups as part of civil society. Second, the liberal interventions made by most pro-control NGOs perpetuate a hierarchical world military order while trying to mitigate its worst excesses. Third, there are a variety of seemingly 'uncivil' actors arrayed against the pro-control NGOs, such as the National Rifle Association in the US, discussed in Chapter Six, and arms capital, discussed in Chapter Three.

Global civil society?

In many accounts of global civil society, NGOs are seen as key actors in an emerging regime of global governance in the era of globalisation. Such accounts are premised on the ending of the Cold War and the ostensible end to ideological conflict that accompanied it, which is taken to mean that governments and international institutions have become more responsive to peace and human rights groups (Kaldor 2003a, 79). One strand of this is the 'cosmopolitan' approach of scholars such as Mary Kaldor and David Held. Kaldor, for example, sees globalisation (understood as intensified global interconnectedness since the end of the Cold War) as having changed the meaning of civil society: it is 'no longer confined to the borders of the territorial state' (Kaldor 2003a, 1). She refers to 'an emerging framework of global governance, what Immanuel Kant described as a universal civil society, in the sense of a cosmopolitan rule of law, guaranteed by a combination of international treaties and institutions' (Kaldor 2003a, 7). Similarly, Held sees global governance emerging along the lines of a 'global social democracy', driven by liberal and social democratic European states, liberals, Southern states, NGOs, social movements and progressive economic forces (Held 2004, 166).

For theorists of global governance such as Jan Aart Scholte meanwhile, global civil society is part of an emerging regulatory regime which, despite various practical shortcomings, is making a positive difference in terms of increasing democratic accountability in global governance. In a world of 'more polycentric governance', in which sub- and suprastate agencies operate alongside national states, civil society organisations direct their energies towards a wider range of governance agencies, beyond the state (Scholte 2004, 214). Under globalisation, we see the emergence of 'supraterritoriality', of 'a realm that substantially transcends the confines of territorial place, territorial distance, and territorial borders', while not fully replacing 'the old geography of territorial spaces' (Scholte 2002, 286). This goes hand in hand with a shift from statist government to multilayered governance, in which civil society actors 'deliberately seek to shape the rules that govern one or the other aspect of social life' (Scholte 2002, 283, 287-8).

Other scholars, meanwhile, are more wary of the claims to globality. Price, for example, argues that civil society is too 'uneven and issue-specific' to be labelled global; he prefers the language of transnational civil society, understood in terms of 'a set of interactions among an imagined community to shape collective life that are not confined to the territorial and institutional spaces of states' (Price 1998, 615). Others emphasise 'the border-crossing nature of the links and the fact that rarely are these ties truly global, in the sense of involving groups and individuals from every part of the world' (Florini and Simmons 2000, 7).

Whether they use the language of global or transnational civil society, such accounts give a pluralist account of the spreading of political authority in a globalising world. The advantage of these accounts is that they do not treat states as unitary actors; fractions within states can ally themselves with non-state actors who share their values and interests. This finds political expression in the claims made about the International Campaign to Ban Landmines, discussed in the Introduction. Indeed, the then UN Secretary General Kofi Annan argued that the 'fruitful cooperation' between the UN and non-state actors on the landmines issue formed 'the embryo of global civil society' (quoted in Brem and Rutherford 2001, 172). However, there remain two significant problems with such a conceptualisation. First, what is widely called 'global' civil society is dominated by Northern-based organisations and funders, and the already-complex relationship between the state and civil society discussed above becomes even more complicated when thinking about the 'global'. Second, the privileging of cosmopolitan agency in such accounts is selectively inclusive and not truly global.

The geography of global civil society is uneven and suggestive of ongoing hierarchical North–South relations. NGOs can be found in their thousands across both North and South, but the concentration of NGO headquarters is firmly in the North. The US is home to the highest number of international NGO headquarters (3,464 out of a world total of 17,968 in 2001), followed by the UK (1,884), Belgium (1,873), France (1,460) and Germany (938) (Anheier and Stares 2002, 318-22). These five states alone are home to over 50 per cent of all international NGO headquarters. Low-income countries have 35 times as many organisation memberships as headquarters, while for high-income countries the ratio is 5 to 1 (calculated on the basis of figures in Anheier and Stares 2002, 318-23). So despite the proliferation of NGOs around the world, the direction and control of international NGO activity is firmly concentrated in Western Europe and North America. This is borne out empirically in the case of NGO activism on the arms trade: there has been a proliferation of NGOs working on arms issues, predominantly small arms issues, but campaign content and direction remain largely driven by European and US-based NGOs.

NGO funding is also Northern-dominated, unsurprisingly. Most of the funding for international NGOs comes from the development industry, led by multilateral agencies, with Western, primarily US, foundations the second largest source of funding (Pinter 2003, 419). Within this, the US state plays an ambivalent role. While it does not give much in the way of formal overseas development aid (ODA), it builds relationships with the South through military aid and also through philanthropy. Philanthropic foundations have acted as an adjunct of US foreign policy, playing a conscious role in building and perpetuating US international hegemony in the post-World War Two era (Parmar 2002, 13-14). The expansion of philanthropy in relation to US foreign activities has served to project US institutional design on to global civil society and further culturally embed socio-economic transformation (Vogel 2006, 636). The role of philanthropic foundations in funding global civil society activism also speaks to the need not to take the state and the market as substantively separate. Although formally private, the large foundations 'are linked not only to large corporate resources but also to the establishment within the US political system' (Parmar 2002, 25). As discussed in Chapter Three, large US foundations such as Ford and MacArthur have been key funders of NGO activity on the arms trade, especially in the 1990s. However, US funding for NGO activism on the arms trade is shaped by the domestic gun rights and War on Terror agendas. Funding for small arms control, the mainstay of international NGO activity on the arms trade, is limited, and US foundations have turned their attention back to nuclear proliferation as a key security challenge since September 2001.

The possibility of a 'global' civil society also reproduces the problematic understanding of its relationship to the state discussed previously. Cosmopolitan democracy theorists such as Held envisage a global polity regulated by cosmopolitan law, distinct from the Westphalian and UN Charter models of democracy (Held 1995). Global governance theorists such as Scholte see the rise of supraterritorial, global spaces alongside the territorial state (Scholte 2002). Such accounts envisage global civil society transcending the states system and bringing with it a more pluralised form of democracy. Although this is deemed a progressive move, what such accounts sideline is the way in which civil society can only be understood in relation to the state, in particular to the modern, capitalist form of state, and that the international system of states is inextricably tied to the global capitalist economy. As Colas argues, the liberal fallacy of reducing capitalist social relations to simply another source of power sidelines the structural features of capitalism. As a result, liberal scholars end up 'fetishizing the political expressions of global capitalism by assuming that the political forms of rule it throws up can be transformed in isolation from the social relations that underpin this system' (Colas 2002, 160). And in political practice, this most often results in the assumption that the worst and most direct excesses of the arms trade – its effects on populations in the South – can be ironed out without transforming the social relations of the world military order that give rise to them. The ways in which this plays out and is contested is detailed in subsequent chapters.

While scholars working within the Marxist tradition are more attentive to the relationship between states and capitalist social relations, they are not agreed on whether we are witnessing the emergence of a global, transnational or international civil society. Rupert, for example, argues that we should not become transfixed on whether there is a 'strict institutional correspondence of state/civil society', but rather should focus on 'the relations of coercion/consent at play in various historical social forms, none of which are wholly understandable in abstraction from their relation to capitalism and its peculiar forms of social organisation'. That is, given the transnational nature of capitalism, we should be open to the intellectual and political possibilities of a global civil society beyond the boundaries of the nation state (Rupert 1998, 431-3; see also Robinson 1996, 381; Cox 1983, 171). Colas, meanwhile, prefers the language of international civil society, as it 'simultaneously undermine[s] and affirm[s] the legitimacy of the modern states-system' (2002, 15). As discussed in Chapter Two, I prefer to think in terms of imperial relations structured around an internationalising state with the US at its centre. Civil society, structured as it is in relation to capitalist social relations and forms of state, is also internationalising, dominated by the US and other Northern powers, and demonstrative of imperial North–South relations. While the internationalisation of civil society is to be expected – and encouraged – from within a Marxist framework, a postcolonial critique draws attention to the asymmetries in its expansion. In this, the South features predominantly as a 'site of Western good intentions, of humanitarian intervention and development assistance' (Barkawi and Laffey 2002, 112) and both the North and South are mutually constituted through NGO activity.

A second problem with mainstream accounts of the global nature of civil society is the privilege given to cosmopolitan forms of agency. In effect, the terms 'global' and 'cosmopolitan' are conflated. Kaldor, for example, uses the terms 'universal', 'global' and 'cosmopolitan' interchangeably (2003a, 2; 2003b). As Mittelman points out in relation to Held's argument, the agents tasked with promoting global social democracy are 'those with superior capacities of awareness, the ability to project cosmopolitan values and norms', in an echo of the 19th century European *mission civilisatrice* (Mittelman 2005, 116; see also Darby 2004, 8). Key actors tasked with promoting and reforming globalisation are the direct inheritors of liberal ideas that facilitated colonial practices, thus obscuring the role of modern liberalism in creating the problems to which these actors now claim simply to respond. The claim that civil society is (in the process of becoming) global thus masks the particularity of the actors that liberals perceive to be populating it.

The defence against critiques that attempt to recover the violent history of liberalism is that 'it is a mistake to throw out the language of equal worth and self-determination because of its contingent association with the historical configurations of Western power' (Held 2004, 156-7). Yet, historically speaking, the emergence of the modern European state system was 'coterminous with, and indissociable from, the genocide of the indigenous peoples of the "new" world, the enslavement of the natives of the African continent, and the colonisation of the societies of Asia' (Krishna 2001, 401). While the urge to imperialism is inherent in liberalism but not a necessary outcome (Mehta 1999), Western liberalism has a bloody history. As Jahn argues, liberalism's 'others' are excluded from liberal norms because the definition of otherness 'is *prior*' to theories of politics and the international (Jahn 2005, 618, emphasis in the original). Held asks us not to dismiss liberalism because it contains the promise that one day, via some of the very actors that are implicated in the emergence of this state of affairs, the world's problems will be solved or at least mitigated. The effect of this is to enact a strategy common to International Relations discourse: the 'eternal deferment of the possibility of overcoming the alienation of international society that commenced in 1492', which serves to justify contemporary and historical violence and inequality because 'the present is inscribed as a transitional phase whose violent and unequal character is expiated on the altar of that which is to come' (Krishna 2001, 402). Liberal accounts of globalisation, global civil society and progressive change thus serve to entrench and perpetuate some of the very practices they claim to oppose. While NGOs are often some of the most vocal critics in world politics, we need to understand the historical context of their emergence and of the values they seek to promote. Liberalism's self-image as tolerant, inclusive and non-violent does not match up to its historical expression. As such, the liberal practices and values promoted by NGOs are also worthy of critical scrutiny.

Global civil society as non-violent

Finally, there is broad agreement within the literature that global civil society is non-violent (Kaldor 2003a, 3; Kaldor *et al.* 2005, 2; Keane 2003, 8, 12; Price 2003, 580). In asserting that global civil society is a non-violent realm, authors draw on the historical link civil society has had with civility and the removal of violence from the public sphere. Kaldor argues that the term assumes 'a rule of law and the relative absence of coercion in human affairs at least within the boundaries of the state' (2003a, 7). She argues that the civilising process was historically based on the establishment of the state's monopoly on violence and taxation. Citing Tilly, she acknowledges that the construction of these monopolies was bound up with wars against other states, through which interstate war became the only legitimate form of organised violence (Kaldor 2003a, 32-3; also 1999, 18, 20). In such a way, processes of internal pacification and external aggression accompanied the rise of capitalism and the nation-state system, meaning that civil society came to be defined as a non-violent sphere.

This narrative of the pacification of the domestic sphere and of the civilising of civil society is not innocent, however. First, the internally pacific, outwardly violent form of state is based on the European historical experience and is only partially or weakly instantiated in the South due to dependent processes of state formation (Bromley 1994). Thus, the very idea of the arms trade as a feature of states' external relations is historically and geographically specific, and plays out differently in the South compared to the North, as we shall see in chapters Five and Six. Second, the emphasis on the emergence of national states in International Relations scholarship and indeed the definition of the subject as inter-*national* relations has the effect of sanitising the violence of world history and bracketing 'questions of theft of land, violence, and slavery' (Krishna 2001, 401, 406). Recovering the violence that has historically been central to the emergence of the nation-state system and has gone hand in hand with discourses of freedom and of tolerance casts a different light on contemporary understandings of the claim to a non-violent global civil society.

It is not that contemporary global civil society authors deny that violence and incivility take place in the world; rather, they exorcise it from view. Keane, for example, locates violence on the edges of civil society: 'On the outskirts of global civil society, and within its nooks and crannies, dastardly things go on, certainly. It provides convenient hideouts for gangsters, war criminals, arms traders and terrorists' (Keane 2003, 12). This assumes that violence is unusual and located on the periphery of world politics, rather than central to capitalist globalisation and the constitution of the state and global civil society. Keane admits that there is a 'stench of violence' surrounding talk of civility, that 'The foundations of civil societies have often been soaked in blood. "Civilised" worldliness typically developed hand in hand with profoundly "uncivil" or barbaric forms of domination' (Keane 2003, 30).

The idea that the violence committed via 'arms traders' is to be found 'on the outskirts of civil society' is unsustainable. Not only has violence historically accompanied the processes of pacification and state formation in the emergence of the modern international capitalist state system, but a more Marxist conception of civil society requires us to consider arms capital to be found squarely within civil society rather than on its edges. The incorporation of NGOs into the US-led War on Terror as 'a force multiplier, such an important part of our combat team' (Powell 2001) further muddies the waters of NGOs' supposed pacific ethos. Similarly, Kaldor's claim that civil society is 'an answer to war' (2003a) is especially interesting when one examines the arms trade, which provides the instruments of war and accompanies the transfer of social ideas and institutions related to organised violence. As such – and against scholars such as Keane who explicitly name NGOs such as Amnesty, Oxfam and Saferworld as undertaking 'civil initiatives against incivility' (Keane 2003, 154) – through their representations of the arms trade, development, human rights and conflict that reproduce imperial hierarchies, NGOs may well be complicit in war and organised violence, rather than an answer to it.

A further issue with regard to violence is the relationships between actors commonly taken to represent civil society. As Pasha and Blaney argue, 'where civil society prescribes a hegemony of liberal civic culture (simultaneously experienced as the domination and/or deculturation of many of its members), the result will be a periodic and irresolvable problem of policing the noncivil within civil society' (1998, 424). This is of interest because it highlights the tensions that can be generated within civil society: the emphasis on particular, liberal models of behaviour generates resistance that must be policed. In addition to questions regarding the partners Northern-based NGOs choose to work with in the South, it also raises the issue of relations between Northern-based groups. As discussed in Chapter Three, there is a degree of policing that goes on within the NGO world that suggests that power relations between NGOs also need to be addressed when considering the effectiveness of the sector as a whole.

In part, attitudes to violence are a question of political strategy and morality. In their attempt to transform a hegemony based on violence, many activists refuse to use physical force; for them, the use of force would further entrench the social phenomenon they seek to challenge. Enacting a prefigurative politics – that is, acting in line with the values one wants to see operationalised across society – is, for many activists, a key principle. But it is also a means of disciplining: direct action protestors who do not eschew violence are further sidelined as political actors. Yet such actors usually have a political rationale for not ruling out the use of physical force and, indeed, part of their action is to challenge dominant conceptions of violence (see e.g. Gordon 2007, ch. 4; Sullivan 2005).

For example, their use of physical force is often conducted against property, challenging the capitalist definition of violence that includes violence against private property. Indeed, based on the violence that has historically been central to the creation of private property, they may not even classify such action as violence. The use of the state's coercive apparatus in the protection of private property can, in turn, further stimulate violent conflict. And there are several examples of anti-arms-trade activists physically disarming military equipment as a means of preventing violence, which can be understood as an act of pacifism and demilitarisation, rather than an act of violence (e.g. Zelter 2004). Both types of activism can serve an instrumental as well as a symbolic function, intervening against the state-sanctioned use of force and attempting to open space for discussion over the social meanings of force, violence and coercion. Yet the liberal emphasis on non-violence further entrenches the hold that the state has on the claim to legitimate violence; the use of force against the state-sanctioned means of violence is still widely socially unacceptable.

This book does not investigate the understandings and strategies of activists who do not rule out the use of physical force. This is because such activists tend to organise themselves in more informal and decentralised ways than NGOs, and the focus of this book is on understanding NGO activity through a re-reading of debates about global civil society. However, as discussed in chapters Three and Seven, debates about violence and non-violence are a live issue for organisations such as CAAT, given their positioning at the more direct action end of the activist spectrum compared to other NGOs.

Conclusion

Civil society is most commonly understood to be a sphere separate from the state and market, progressive, increasingly global, and non-violent. Reading the mainstream literature against postcolonial and Marxist critiques is a counter to the pervasive liberalism and Eurocentrism of such accounts. It reminds us that the separation of state from civil society is methodological rather than substantive, and grounded in the modern European historical experience. This means that the ambiguity of civil society should be at the forefront of analysis, in particular its role in both buttressing and challenging hegemony, and the centrality of violence to civil society and state formation. We must also problematise the universalism inherent in the application of the concept beyond the European world in the celebratory guise of the (potentially) global nature of civil society. Not only is it dominated by Northern organisations, but the very possibility of a 'global' civil society in an internationalising capitalist nation-state system must be questioned. Further, the emphasis on cosmopolitan forms of agency is not globally inclusive, relegating as it does 'uncivil', non-liberal and anti-capitalist forms of agency to the margins, leaving them open to disciplinary and interventionary action and, at times, force. This is suggestive of the imperial nature of North–South relations and mutual constitution of both categories.

Overall, the concept of global civil society carries significant ideological weight in excess of its analytical utility. The concept of civil society, in being emptied of one of its potential meanings (that of the perpetuation of hegemony), 'has been appropriated by those who foresee an emancipatory role for civil society' (Cox 1999, 10). Understood primarily in terms of its prophetic function, global civil society 'seems to have been studied into existence by scholars who self-consciously have blended analytical and normative concerns in order to justify *their* particular vision of a global community' and thus, 'being expressions of the will to govern, theories of global civil society are part of the problem rather than of the solution' (Bartelson 2006, 374, 389, emphasis in original; also Duffield 2002a, 1053).

Re-reading NGOs as global civil society actors in this light, while avoiding committing the liberal fallacy of *equating* NGOs with global civil society, prompts us to reconsider their activity and is the task of the remainder this book. The chapters that follow examine NGO activity in relation to the various trends in play in the arms trade, asking how NGOs understand and campaign on a variety of arms trade issues, what the relationship is between their arguments and strategies, and what effects their activity has in terms of the production, reproduction and/or challenging of international relations.

2 • What's the Problem?
NGOs and the Arms Trade

The arms trade generates concern from a variety of corners, whether it is understood as a guarantor of state security and national defence or as a threat to human security. Million-dollar state-sanctioned arms deals raise allegations of corruption and pork-barrel politicking, while the illicit trade in small arms is hailed as a key feature of the post-Cold War terrain of conflict, contributing to insecurity, underdevelopment and human rights violations. NGOs have become key agents in the battle over the construction of the arms trade as an issue in international relations. Engaging in mass campaigns, media work, policy development, lobbying and advocacy, they try to create awareness of the arms trade and shift popular and elite understandings of it. This chapter examines the NGOs' overarching representations of the arms trade as a problem in international politics. For example, is the problem one of the illicit trade in small arms, or the state-sanctioned trade in major conventional weaponry? Is globalisation undermining states' abilities to control the trade, or are states willing accomplices in these transformations? Should NGOs try to reform the arms trade or abolish it?

There are two main visions of the arms trade put forward from the NGO community: one reformist, the other transformist. Reformists 'wish to correct what they see as flaws in existing regimes while leaving underlying social structures intact' (Scholte 2002, 284). This approach, exemplified by Amnesty, Oxfam and Saferworld, accepts the legitimacy of the arms trade and seeks tighter regulation within existing parameters, through the improvement of laws and policies, better implementation, and increased transparency. The main problem posed by the arms trade is deemed to be the unregulated trade in and to the South, predominantly that of small arms, which poses a threat to security and development and increases the likelihood of conflict. Transformists, meanwhile, 'aim for a comprehensive change of the social order (whether in a progressive or a reactionary fashion)' (Scholte 2002, 284). CAAT and BASIC both operate with transformist visions. CAAT has an abolitionist stance towards the arms trade and focuses on the relationship between the state and arms capital as the driving force behind arms exports that contradict the government's publicly stated policies. BASIC, meanwhile, accepts the legitimacy of the international arms trade but starts from a position of critiquing Western military predominance in an over-militarised international security arena. To a degree, the two share a similar vision of international security, although there are no formal links between them.

Reformists and transformists

A key point of difference within the NGO community is their overall stance towards the arms trade. Amnesty, BASIC, IANSA, International Alert, Oxfam and Saferworld all recognise the international arms trade as legitimate on the basis of states' right to self-defence under Article 51 of the UN Charter. One of the basic principles of the Control Arms campaign for an Arms Trade Treaty is that 'most arms can have a legitimate use under international law for national self-defence and law enforcement', however, they must be strictly controlled so as 'to prevent weapons falling into the hands of those who would use them for serious abuses and unlawful acts' (Control Arms 2008a). The Arms Trade Treaty 'would not impede legal international transfers of arms'; rather, it would 'set clear parameters for legal international transfers of arms and would assist the identification and elimination of illicit transfers' (Saferworld 2009a, 2). The work of these NGOs thus revolves around tighter regulation of the international trade. They focus on eradicating the illicit trade and improving regulation of the legal trade so as to abide by national and regional control regimes, and are pushing for an international control regime.

Within this shared framework, the various NGOs conceptualise the problems posed by the arms trade in different, yet complementary ways. For Amnesty International, the arms trade is a human rights issue when arms transfers are used to commit or facilitate human rights violations or abuses. Beyond that, Amnesty claims not to take a position on the arms trade *per se* or on economic sanctions or punitive measures directed against any state. As described by an International Secretariat researcher, Amnesty opposes the transfer of military, security and police equipment when this can reasonably be assumed to contribute to human rights violations within Amnesty's mandate, such as unlawful killings and torture. Key targets of Amnesty campaigning are thus military equipment used in torture, weapons that have indiscriminate effects, and weapons that can be linked to specific human rights violations, according to a campaigner from the UK section. So in practice, Amnesty often focuses on particular categories of conventional weaponry such as small arms because of the frequency with which they are used in human rights abuses and violations of humanitarian law, as part of an overall attempt to make it more difficult for those who abuse human rights to acquire the equipment to do.

Just as Amnesty is concerned with the arms trade as it relates to human rights, Oxfam works on arms trade issues in relation to development concerns. Its concern with the arms trade comes from its operational experience in its mission to find lasting solutions to poverty, suffering and injustice. It argues that 'easy access to arms increases the levels of human suffering', regardless of the history and causes of the conflicts in the countries in which it has been working for the past sixty years (Oxfam 2002b, 3). It is focused predominantly on small arms: its work in the field has shown it that 'conflicts are fuelled by the international transfer of arms, most notably small arms and ammunition' (Oxfam 2001, 3). Addressing the demand for small arms, which includes factors such as 'poverty, insecurity, lack of sustainable livelihoods, lack of equitable access to services, assets and opportunities' is therefore an important part of its mission (Oxfam 2001, 3).

Small arms are also the main focus of International Alert and IANSA. International Alert has an overarching mission of peacebuilding. It has been working on small arms issues since 1994, when it 'identified unregulated small arms proliferation and misuse as one of the world's most pressing security issues' (International Alert 2006a). It understands small arms as a 'barrier to peace': they are cheap, easily obtainable, easy to use, are widely held by non-state actors, and used in conflict and crime. International Alert thus aims for the removal of weapons from societies as part of promoting peace (International Alert 2006a). IANSA, meanwhile, focuses on small arms proliferation as a problem of gun violence, which impacts negatively on human rights and human security. Its overarching, twin objectives are stronger regulation on guns in domestic societies and better controls on arms exports in pursuit of policies that protect human security (IANSA no date, a).

For Saferworld also, arms export controls are part of a wider human security agenda: 'Irresponsible arms exports have a massive human impact and can contribute to internal repression, undermine development, facilitate human rights abuses, fuel and sustain conflict' (Saferworld no date, a). At the national level it works 'to ensure that the UK's arms export rules are effective and rigorous' (Saferworld no date, b). It sees controversial arms exports as the 'missing link' in UK foreign policy, a contradiction in an otherwise benevolent foreign and development policy (Mepham and Eavis 2002). While it points to the government's relationship with arms companies as a key reason why it makes 'ugly arms export decisions' (Saferworld 2007a), this is not the focus of its work. As well as work on conflict-sensitive development and security and justice sector reform, it works on small arms issues in the South, premised on the argument that their proliferation fuels violent conflict and undermines development.

Since the early 2000s, the Control Arms NGOs, predominantly Amnesty, Oxfam and Saferworld, have paid increasing attention to the globalisation of the arms trade. They emphasise the increasingly internationalised character of arms production, such as the growth of international sourcing of components, the flourishing of brokers and dealers, increased trading in technology, the production of weapons across multiple locations, and the blurring of the distinction between military and civilian production and trade (Control Arms 2006b). These processes, part of a post-Cold War shift precipitated by a decline in military spending and procurement, find their counterpart in an increased export drive, 'often regardless of the ethical consequences' (Amnesty International 2006, 5). As a result of these processes, the NGOs argue that 'no country has an autonomous arms industry' (Control Arms 2006b, 8). Industrialised states remain the world's largest arms exporters, but key Southern states such as Brazil, China, India, Israel, Pakistan, Singapore, South Korea and South Africa are emerging as significant exporters. Military production in these states is often oriented towards export, and whilst 'it is legitimate for these countries to seek to develop their industries and compete with the traditional manufacturers' and 'to compete for an increasing share in the global arms market as they do in other manufacturing sectors', they must ensure that their exports are in line with their responsibilities under international law (Control Arms 2006b, 10). Yet among these emerging players, 'national arms export controls vary, and do not always include explicit criteria or guidelines for authorising arms transfers that fully reflect states' existing obligations under international law', even though it is in states' interests both to ensure a level playing field and to protect human rights (Control Arms 2006b, 3-4). In response to the impact of globalisation, the NGOs are campaigning for an Arms Trade Treaty to institutionalise states' responsibilities to regulate the trade.

A key element of the globalisation of the arms trade has been the blurring of the boundary between the public and the private, which, for the Control Arms NGOs leads to a practical focus on issues such as brokering, transportation and marking and tracing of weaponry. Amnesty argues that 'Growing state-sponsored out-sourcing and the increasing private mediation of international arms distribution and procurement is adding to the risk of arms being delivered, diverted and used for grave human rights violations' (Amnesty International 2006, 3). Globalisation has led to increased competition for exports: there was an 'export rush' at the end of the Cold War, which has meant that 'arms trade routes are becoming more complex, requiring even more differentiated logistical, transport, brokerage, and financial arrangements' (Amnesty International 2006, 3). The end of the Soviet bloc 'resulted in large and loosely controlled stockpiles of conventional weapons being offered for sale on the international market' (Amnesty International 2006, 5), and the post-Cold War international environment features growing South–South arms routes and transfers, and increased outsourcing of defence logistics, especially by the US. The latter process features a growing role for commercial agents, with military and civilian goods often mixed together, and evidence that some transporters will deliver humanitarian aid one week and arms supplies the next, including in sanctions-busting deliveries (Amnesty International 2006, 49). As a result, 'As the chain of sub-contractors is extended internationally, the role of arms brokering, transport and logistics can become ever more remote from democratic oversight and accountability' (Amnesty International 2006, 44). The growing complexity of supply chains has obscured states' responsibilities, and national legal and regulatory frameworks are inadequate. Again, this reinforces the NGOs' call for an international, legally binding Arms Trade Treaty.

In contrast to this reformist vision, which seeks tighter control of the trade through national, regional and international regulatory frameworks, CAAT and BASIC adopt transformist approaches, which operate outside of the existing mainstream frame of reference. CAAT's overarching goal is the reduction and ultimate abolition of the international arms trade, because of its role in undermining human rights, security and economic development locally, regionally and globally. It allies this with the goal of progressive demilitarisation within arms-producing countries. As steps towards these goals, it calls for: an end to government political and financial support for arms exports; an end to exports to oppressive regimes and countries involved in armed conflict or regions of tension, or in which social welfare is threatened by military spending; and the promotion of policies to orientate the UK economy towards civil production (CAAT no date, a).

Rather than focusing on the illicit trade, CAAT's concern is the legal, state-sanctioned trade, in particular that of the UK. It argues that arms companies 'wield immense influence and political power' in government as a result of their privileged access, which 'undermines democracy' by turning government policy into 'arms company wish lists' (CAAT, no date, b). The web of relationships between companies and government reinforces a 'militaristic approach to security and problem solving within government that makes it extremely receptive to the approaches of arms companies', and a more general pandering to the corporate agenda (CAAT 2006a). CAAT retains a focus on nation states and the relationship between states and arms capital, making demands that can primarily be met at the level of the UK state. To the extent that it discusses the globalisation of the arms trade, it does so to challenge arms companies' claims to benefit the national interest. For example, it regularly challenges BAE Systems' claims to contribute to the UK economy, arguing that both the internationalised nature of arms production and the search for lower production costs and higher profit margins mean BAE's commitment to the UK is purely tactical (CAAT 2009a; Stearman 2009).

BASIC's work on the arms trade starts from the supposition that 'international security policy is over-militarised'; Western and international policy should be driven by 'collective security and disarmament'. BASIC's organisational focus is on the West, as the US and UK 'are the most influential but least critiqued entities in global politics' (BASIC 2001a). It emphasises the predominant role of the UK and, in particular, the US in the manufacture and trade in conventional weapons, which are the most commonly used category of weapons in contemporary conflict. BASIC argues that sales of major conventional weaponry and small arms alike are 'currently subject to insufficient coordination and control. Arms exporters exercise little restraint over their own sales, and participate in only vague and non-binding international cooperation' (BASIC no date, a). A key goal for BASIC is tighter and more harmonised US and EU regulation of the arms trade, which is in line with its organisational remit as a transatlantic security NGO.

BASIC seeks to encourage responsible transfers rather than call for an end to the arms trade, according to a senior staff member, on the basis that denying transfers would encourage indigenous weapons production in non-producing or small-scale states. However, there is a potential affinity between its position and CAAT's in that they both emphasise the disproportionate role of the major capitalist powers and military spenders in the global arms trade, are critical of the relationship between arms companies and the UK state, and of UK military posture. As discussed in Chapter Three, however, the strategies they use to generate change are very different. In addition, BASIC has moved away from conventional arms control issues in recent years, returning its focus to nuclear disarmament in 2007. It argues that nuclear weapons are militarily ineffective against terrorism and inappropriate as a response to the non-military nature of many future threats (BASIC 2006a, Ev. 111). The nuclear weapons debate is largely framed in terms of terrorism and non-proliferation, making it yet another imperial project that Russia, China, and India will not buy into, according to a senior staff member. As such, the international debate around nuclear weapons needs to be reframed.

CAAT and BASIC occupy some common ground in their understandings of the nature of international relations in the early 21st century, as evidenced by reports written by the same researcher. They emphasise the dominance of US corporations in global defence markets and US technological and industrial supremacy (Schofield 2006). This translates into direct military advantage but also 'reinforces a hierarchy of strategic power and influence through which the United States can bring pressure to bear on other countries for broader strategic support' (Schofield 2008, 9). The UK is the starkest example of this, in that it is tied to the US as a junior partner, while Europe more broadly has a more ambivalent role: European restructuring is an attempt to 'create a globally competitive industry that could challenge the major American corporations' (Schofield 2008, 7).

Both NGOs emphasise the relationship between arms capital and the state. As well as CAAT's overall campaign focus of recent years, BASIC refers to 'the growing stranglehold of BAE on procurement' in the UK (Schofield 2006). The new technologies and forms of military production associated with the post-Cold War threat agenda are 'additional to, rather than at the expense of, traditional military preparations and procurement' (Schofield 2008, 10). That is, 'demand for traditional force projection' will continue, 'not least because of the vested institutional power of vested military–industrial interests, allied to additional capabilities for anti-terrorist operations' (Schofield 2008, 11). In contrast to the Control Arms NGOs, which paint a picture of diffusing, networked power, BASIC and CAAT locate power in US predominance and state–capital relations.

Overall, there are a variety of objectives within the NGO community. The Control Arms NGOs understand the problem of the arms trade to be unregulated exports to sensitive regions or countries. With improved policies and tighter implementation, these anomalies could be ironed out. In addition, there is a practical focus on small arms proliferation in the South, shared by IANSA and International Alert, as this is the most pressing aspect of the arms trade: the NGOs focus on where the effects are worst felt. That is, people die, are maimed, and have their livelihoods destroyed by small arms every day – as indicated by a counter at the top of the IANSA website documenting the number of gun deaths since 1 January 2009. CAAT and BASIC, while they also largely focus on controversial exports, locate the impetus to such practices elsewhere, namely in the relationship between the arms industry and government and the geo-political ramifications of US predominance. Overall, the regulatory approach of the Control Arms NGOs is a reformist position, while the abolitionist approach of CAAT is transformist, with BASIC oscillating between the two. To assess the adequacy of these visions, we need first to examine the role of military production and trade in international relations more closely.

Military production and trade in the post–Cold War era

Military production and trade are dominated by the US, Russia and Western European states. In the period 2004–8, the US and Russia accounted for 31 per cent and 25 per cent of international arms transfers, respectively. Taken together, EU member states accounted for 34 per cent, with the UK, France and Germany the largest exporters within the bloc (Wezeman et al. 2009). Asia, Europe and the Middle East remain the main recipient regions in the post–Cold War world. While there was a shift away from the enormous oil boom arms contracts with the Middle East of the Cold War years, the War on Terror has seen a resurgent emphasis on military aid and arms sales, reversing the contraction of the international arms market that accompanied the end of the Cold War. US exports are mainly to the Middle East, Asia Pacific and European NATO members. Russian exports are predominantly to China and India, which were the world's largest arms importers in 2004–8, accounting for 11 per cent and 7 per cent of the world's arms imports respectively. However, exports to China in 2008 were at their lowest since 1999 and are unlikely to rise again as contracts have been fulfilled and no new ones have been announced. Most arms transfers from EU member states went to other EU states, Asia and the Middle East, with India and the US as important and growing markets (Wezeman et al. 2009).

The international arms trade is predominantly a North–South phenomenon: approximately two-thirds of all international arms transfers are to developing states (Grimmett 2008). While South–North and South–South transfers are increasing, due to the internationalisation of military production, the growing specialisation of some Southern producers and their role in transferring small arms, the bulk of the remaining third of all international exports are intra-Northern. But international arms exports are just one element of the arms trade: domestic sales from companies to their national governments are also part of the arms trade. When we factor this in, the arms trade looks rather different. World military spending as a whole increased by 45 per cent in real terms over the period 1998–2008. The US accounted for 58 per cent of this increase and by 2009 was responsible for 42 per cent of the world's total military spending, dwarfing that of the next biggest spenders (China, France, the UK and Russia, with 4–6 per cent of the world's total each) (Perlo-Freeman *et al.* 2009, 179, 183). There were considerable increases in US military spending under the Bush administration, to the point where it was higher in real terms than at any point since the end of World War Two (Perlo-Freeman *et al.* 2009, 184). And within the EU, the establishment of the European Defence Agency (EDA) and the creation of EU-wide arms export practices are aimed at strengthening the defence industrial base, with the EDA calling for increased military expenditure in European states, to be spent on procurement (e.g. European Defence Agency 2006).

There has also been a shift in the nature of military production, enhanced by the end of the Cold War. Since the advent of industrialisation and in particular with World War Two, military production within the North has become increasingly capital-intensive and technologically sophisticated. Since the end of the Cold War the arms trade has seen a shift away from complete platforms, such as tanks, fighter aircraft and naval vessels, towards electronic systems, information technology and components, with a concomitant increased role of 'spin-in' to the military sphere from traditionally civilian industries (Dunne and Surry 2006, 414-15). However, this putative 'Revolution in Military Affairs' has not overcome the bureaucracy, waste and gold-plating associated with military production during the Cold War (Kaldor 1983; Hartung 2008).

Small arms[1] production is considerably more diverse and widely spread around the world, due to the lower technological barriers to production of mature and widely available technologies. Small arms are produced in more than 90 countries by at least 1,249 companies (Small Arms Survey 2004, 7).

The most recent figures suggest that in 2006 the global authorised trade in small arms was worth approximately US$1.58bn, while the undocumented trade was likely to be at least US$100m (Small Arms Survey 2009, 7). The top 15 exporters (the US, Italy, Germany, Belgium, Brazil, Austria, the UK, Japan, Canada, Switzerland, Spain, the Russian Federation, the Czech Republic, France and Turkey) were responsible for 83 per cent of all exports between 2000 and 2006, with the US consistently both the top exporter and the top importer (Small Arms Survey 2009, 12).

Of the 875 million firearms (at least) in circulation around the globe, 650 million (or 74 per cent) are in the hands of civilians, 270 million of these in the US (Small Arms Survey 2009, 178-9). As the Small Arms Survey argues, 'A large proportion of the discussion of the global trade in small arms and light weapons is actually a discussion of the United States' (2009, 25). A further 200 million (23 per cent) are held by state militaries and 26 million (3 per cent) by law enforcement officials around the world. Less than 1 million are held by insurgents (figures from Small Arms Survey 2009, 178-9). Despite making up only a tiny proportion of global small arms possession, it is the weapons held by insurgents and non-state forces that are deemed 'most likely to be used to harm' and to be the most destabilising (Small Arms Survey 2001, 2, 77). According to the Small Arms Survey, 'It is the illicit trade in small arms, more than any other aspect of the global arms business, that exacerbates civil conflict, corruption, crime, and random acts of violence' (2001, 165; also Lumpe *et al.* 2000, 2). Disarmament, demobilisation and reintegration (DDR) of non-state armed groups is 'far and away the most visible and best-funded' aspect of small arms disarmament (Small Arms Survey 2009, 180).

Imperial hierarchies in the world military order

The various trends in the arms trade – high-technology intra-Northern production and trade, North–South transfers, and the proliferation of small arms – are points on a spectrum, nodes in an internationalising arms dynamic that contributes to a wider military order that is marked by North–South hierarchy. The arms dynamic refers to the production and spread of military technologies (Held *et al.* 1999, 89; Buzan and Herring 1998), which means we need to relate international arms exports to broader patterns of military production and the spread of military technology. The world military order refers to the structure and dynamics of military activity as a site of interaction in its own right, albeit one shaped by and shaping the wider social structures of which it is a part (Held *et al.* 1999, 89; Kaldor and Eide 1979; Øberg 1980). As Martin Shaw argues in relation to war more widely, while 'It seems difficult to conceptualise Western and non-Western warfare together in a single frame...there is an overriding case for doing so', in that 'A sharp distinction between Western and non-Western wars or warfare is...impossible to sustain' (Shaw 2005, 47). The same can be said of military production and trade: the processes of high-technology, intra-Northern transfers, North–South transfers, and the proliferation of small arms, are different articulations of the world military order. This means that different technological forms and modes of organised violence around the world should be studied within a single analytical frame, and their relationship to each other analysed.

Military production and trade are highly stratified. Scholars differ as to which states populate the various levels or tiers of the hierarchy, but are agreed that qualitative superiority and quantitative dominance are what differentiate them. For example, according to Krause, first-tier producers are those that can undertake technological innovation and produce weapons systems for all military applications, the 'critical innovators'. Second-tier producers can satisfy some of their own needs, as 'adapters and modifiers'; while the third tier comprises those states that attempt to create an indigenous capacity through the import of technology, copying and reproduction of existing technologies (Bitzinger 2003, 88; Krause 1995, 26-32). This and other models of the arms production hierarchy (Bitzinger 2003; Buzan and Herring 1998, 42-6; Neuman 1984; Ross 1989) are valuable in helping us map the world military order. While his conceptualisation is realist in orientation – assuming an anarchic system of sovereign states – Bitzinger hints at a different way of thinking about this hierarchy with his description of the arms production system as 'a huge "hub-and-spoke" model: a few large first-tier firms at the core, serving as "centres of excellence" for weapons design, development and systems integration, with global supply chains extending out to second-tier states on the periphery' (Bitzinger 2003, 7).

Taking the core–periphery, or North–South, idea seriously allows us to adapt existing accounts and understand the arms trade as demonstrating imperial hierarchies. The US is the sole first-tier member, due to its unparalleled military budget and predominance in the area of high tech-nology. It also exercises strict controls over foreign ownership and access to its military technologies and defence market. While even the US is not fully autonomous or self-sufficient in military production, any internationalisation is tightly controlled by the US state. For example, when military equipment is procured from European companies, much of it is produced in the US, in cooperation with a US company, under US designation and with a high US content (Wezeman *et al.* 2009, 307). And US arms companies bought up by foreign firms must have a separate board made up of US citizens (Perlo-Freeman 2009a, 276). Only three foreign companies appear in the list of the 100 largest recipients of US Department of Defense direct contracts for 2006 (BAE Systems, Rolls Royce and Thales), and most contracts seem to come from the companies' US operations (Perlo-Freeman 2009a, 275). Thus, even in the instances when the US imports weaponry, there is already a strong US connection and the US state controls it tightly. UK-based companies have led the way in trying to secure contracts that accompany growing US military spending: US restrictions on foreign ownership of the defence-industrial base have led to companies, in particular BAE Systems, buying up smaller US companies in an effort to qualify as a US company. In 2007, 41 per cent of BAE Systems' sales were to the US and Canada, while the UK, the company's next largest customer by geographical location, accounted for 22 per cent of sales (BAE Systems 2008a, 100). The UK has a privileged position in relation to the US, with significant levels of collaborative procurement and research and development programmes. However, there is still significant suspicion of the UK in the US, leading to restrictions on technology transfer, making the relationship highly asymmetrical.

On paper, the US has one of the strongest control regimes in the world, but its orientation and interpretation is heavily coloured by foreign policy concerns. Its control regime also applies extra-territorially, with strict controls over issues such as the re-export of military equipment. The US itself, meanwhile, refuses to be bound by other states' regulations and also has a history of engaging in covert transfers and training for foreign policy reasons, including via proxies (Chomsky and Herman 1979; Barkawi 2001; Blakeley 2009). As argued by an NGO researcher, the strict and extra-territorial application of US law in relation to arms transfer activity, including brokering, shipping and financing and so on, means that in Africa, for example, there tend to be fewer private US citizens engaged in illicit trafficking than nationals from Western and Eastern Europe, Israel, the Middle East and South America, whose home countries do not have laws that would catch them. While such traffickers engage in unauthorised transfers, US transfers have occurred under the remit of covert state-sanctioned practices such as CIA interventions (quoted in Garcia 2006, 231). The US also takes responsibility for ensuring the smooth running of the wider arms trading and capitalist system, with the US Department of Justice chasing BAE Systems about allegations of bribery in relation to its arms deals with Saudi Arabia, after the UK government called off its investigation (Reisinger 2008). The US thus demonstrates a combination of national and internationalised state power.

Coercive protection for the internationalising capitalist system has been dominated in the post-1945 era by the US state, which 'houses and exercises direct control over the principal military machine in the world' (Robinson 2004, 138-9), even if its 'dependence on the wider framework of Western and global power networks has increased' (Shaw 1997, 511). That is, empire is not reducible to US state power, but is better understood as 'an internationalising state dominated by the US' (Barkawi and Laffey 2002, 124). This dominance is particularly pronounced in the military sphere, although one of the peculiarities of warfare is the ability of quantitatively and/or qualitatively less well armed parties to exact unexpected victories. US empire involves the integration of other capitalist powers 'into an effective system of coordination under its aegis', institutionalised militarily through NATO and the 'hub-and-spokes networks binding each of the other leading capitalist states to the intelligence and security apparatuses of the US' (Panitch and Gindin 2003, 13, 15). This institutionalisation of internationalisation, as seen through US insistence on collective security, was central to the post-World War Two order (Mabee 2009, 34). In its relations with Europe, the US has not been reliant on direct political-military occupation and coercion in the way that the USSR was because the capitalist nature of social relations in the US allowed it to use non-coercive forms of domination (Saull 2005). Thus, the USSR exercised a greater degree of direct control over military production and doctrine in the Warsaw Pact than the US has exercised over NATO states. In the South, however, US and Soviet relations with proxy forces displayed several key similarities (discussed below).

The second tier comprises the major arms-producing countries of Western Europe and Russia. West European states have been integrated into a US-led internationalising state, although this is not without its tensions, such as the potentially conflicting trends of transatlantic and European integration. The UK government has tried to control processes of Europeanisation to ensure they do not threaten the transatlantic relationship, insisting, for example, that the first head of the EDA was British (see also Dover 2007a, chapter 7). The US, in turn, wants to ensure the EDA does not allow the EU to unify and consolidate its defence industrial policy and thus potentially challenge the US. European states articulate internationalisation as in the national

interest: the UK's 2002 Defence Industrial Strategy, for example, stated
that the UK defence industry should 'be defined in terms of where the
technology is created, where the skills and the intellectual property reside,
where jobs are created and sustained, and where the investment is made'
(MoD 2002, 9), meaning that French-owned Thales is now considered
part of the UK's defence industrial base, for example (O'Connell 2005).
UK government embassy officials and defence attachés expend considerable
energy promoting the interests of arms companies abroad; whether they are
UK- or foreign-owned, if they create employment in the UK they are
deemed to be a UK company (Dover 2007b, 689-90).

Russia remains one of the world's largest arms exporters, primarily due
to its relations with India and China, which account for 20 per cent and
42 per cent of Russian arms exports, respectively (Wezeman et al. 2009,
304). With the restructuring of the Russian military establishment and arms
industry after the Cold War, the latter has become increasingly reliant on
exports, partly because of lack of funding or policy to restructure the defence
industry and stimulate conversion (Berryman 2000, 94). The weakening of
state control under Gorbachev and in the reorganisation of Russian military
forces after the dissolution of the USSR meant that Russian military and
security organisations were 'the primary initial sources of illicit supplies of
conventional arms' in the late 1980s and 1990s, and illicit sales from Soviet
troops withdrawing from Eastern and Central Europe contributed further
(Berryman 2000, 87).

Soviet and US policy towards the South shared some fundamental
characteristics during the Cold War, such as the transfer of military equipment along ideological lines, the related military training, emulation of
military organisation and standardisation of weapons, and the use of proxies
to intervene in conflicts according to ideological and political-economic
imperatives, as well as the establishment of the illicit arms transfer routes
that are still used in today's supposedly 'new' conflicts over resources, gems
and drugs, such as in Angola, Afghanistan and Latin America. One of the
differences between superpower involvement in the South was Soviet
willingness to supply large quantities of weapons that were more suited to
the type of conflict occurring in the South, and to do so rapidly and at low
cost, in contrast to the high-tech specifications of weapons often offered by
the US (Krause 1995, 116).

The third tier of arms producers consists of those industrialised countries
with small but sophisticated industries such as Australia, Canada and Japan,
and newly industrialised states with often niche, but growing military

production capabilities, such as Israel, South Africa, Brazil and South Korea. This segment of the third tier is mostly integrated into the US-led internationalising state. China and India, meanwhile, other members of the third tier, have an ambivalent orientation, with India trying to play Russia off against the US, and China under an arms embargo that the US is keen to keep, despite Western European states' desire to lift it. The fourth tier comprises those states with limited and generally low-technology capability, such as Egypt and Mexico. Southern states that have any significant military production capability occupy the third and fourth tiers, while the first and second are dominated by the North.

Overall, the contemporary arms dynamic is imperial because it perpetuates hierarchical North–South relations, through both Northern integration under US aegis and the ongoing North–South orientation of the international trade. This is not to argue that *all* arms transfers are necessarily imperial, nor that *only* imperial relations are fostered by the arms trade: imperial relations are not the only form of international relations (Barkawi and Laffey 2002, 114). Rather, it is to argue that the imperial characteristics of the arms trade are under-appreciated, in that the North–South hierarchy and mutual constitution that it facilitates do not receive adequate attention.

It has not always been this way. A globalising, industrialised arms industry first started to emerge in the mid 19th century as part of the initial industrialisation of warfare (McNeill 1982, 241), but military technologies and techniques are one of many 'resource portfolios' that have historically diffused from East to West, enabling the rise of the latter to its current dominant position (Hobson 2004). For example, whilst developments such as gunpowder, the gun and the cannon are often hailed as technological breakthroughs in the first major military revolution deemed to have taken place in Europe between 1550 and 1660, these technologies were first invented in China around nine centuries earlier (Hobson 2004, 58-9). And European sailing prowess came in part from the incorporation of Arab inventions, made possible through contact during the Crusades. The European world's present-day military predominance thus came about in interaction with non-Europeans (Barkawi 2006, 51). Processes of transmission, appropriation, diffusion and assimilation are central to the history of military (and other forms of) development, the latest historical phase of which is marked by North–South hierarchies.

A key feature of the contemporary arms trade hierarchy is a capital- and technology-intensive 'global military culture' that favours the acquisition of large, expensive conventional weaponry in pursuit of a professional, modern army (Wendt and Barnett 1993). The concept of a global military culture allows us to move beyond the idea that arms acquisition, whether through domestic procurement or international transfers, is an objective response to military need (for a variety of ways of thinking about this see Buzan and Herring 1998, ch. 11; Mutimer 2000; Eyre and Suchman 1996). Within the North the post-World War Two era has seen a structural shift from labour- to capital-intensive militaries, a shift which has also seen the rise of 'the cult of the weapon-system' in which 'traditional militarism was replaced by "armament culture"' (Shaw 2005, 37; Luckham 1984). And arms transfers, accompanied as they are by personnel exchanges and the transfer of ideas about military tactics and doctrine, encourage the spread of particular modes of industrialisation and military doctrine (Albrecht and Kaldor 1979, 3; Krause 1995, 16). While this applies to the North as much as to the South, it is most pronounced as a North–South phenomenon. Thus, the newly independent states that emerged from the decolonisation process chased the acquisition of modern armaments: by the mid to late 1960s jet fighters had become a defining symbol of their statehood, irrespective of their ability to operate them (Phythian 2000a, 10; van der Westhuizen 2005, 287). As Krause argues, 'links with external powers have shaped the pattern of military development of postcolonial states, and have helped incorporate those states into a global military system' (1996, 333).

There is a fundamental asymmetry in the global military culture, a 'mostly one-way process shaping Third World military development in ways different than would be the case in its absence' (Wendt and Barnett 1993, 337). Supplier states play a heavy role in actively creating demand, through weapon development, by marketing weapons around a global network of exhibitions, and by selling to regional rivals (Phythian 2000a, 35). And their own pursuit of increasingly sophisticated weaponry for their own armies contributes to the valorising of capital-intensive militarisation strategies that states on the periphery have to follow if they are to remain in the club of 'modern' states. In addition, dominant supplier states have typically given access to weapons and technologies that encourage dependent and capital-intensive militarisation (Wendt and Barnett 1993, 336), which further binds Southern states into the world military order.

While Southern elites are active agents in these processes, they act under conditions of hierarchy. Both dominant and subordinate agents are shaped by the interactions between them: they are mutually constituted through their involvement in the arms trade. For example, the sale of Hawk jets to

India by the UK demonstrates not only the mimicking of a former colonial power by an ex-colony, but also impacts on the strategic orientation of the UK military, as the Royal Air Force (RAF) was forced to buy and operate Hawks itself, so as to promote the chances of exporting them, despite military leaders preferring a different aircraft. And arms exports to Indonesia have not only significantly shaped the conflicts between Jakarta and the regions, but the backlash over the role of arms exports in East Timor in the late 1990s coincided with the announcement of an 'ethical dimension' to UK foreign policy and was significant in tarnishing the UK government's reputation and encouraging the quiet death of the ethical tag.

At the other end of the technological spectrum of the world military order is the spread of small arms. Cheap, easy to manufacture, purchase and use, whether legally or illegally, widely available as leftovers of previous conflicts, and easily transferred across borders, small arms are used in armed conflict, crime and domestic violence around the world. As recognised in the earlier days of transnational non-governmental activism in the late 1990s, however, there are 'difficulties in conceptualising an overarching framework for activities to combat the worldwide proliferation of small arms' (Garcia 2006, 50). That is, small arms are the technology of choice in a variety of social relations of organised and individual violence. Small arms are widely used in homicide, suicide, insurgency, armed crime, armed violence and so on, each of which plays out differently in the North and South, as well as within regions and countries.

The end of the Cold War led to a change in the pattern of the trade in small arms but also a shift in the way that international security was conceptualised, which meant that small arms issues became a matter of increased concern. The break-up of the USSR and former Yugoslavia, and diminished superpower control over Southern proxies as well as NATO and Warsaw Pact stockpiles meant that the number of suppliers of small arms increased significantly, with the effect of releasing increasing numbers of small arms on to the international market; this combined with a changed threat agenda that focused on the spread of ethnic, tribal and religious conflict, and the privatisation of authority and violence (Klare, cited in Garcia 2006, 41). As a result, some scholars tried to reframe arms control, focusing in particular on the category of small arms and emphasising human security in the context of civil conflicts and the so-called new wars (Boutwell *et al.* 1995; Boutwell and Klare 1999; Singh 1995; on the recent history of academic research into small arms, see Garcia 2006, 35-42). Human security emerged as a key post-Cold War concept, emphasising freedom from fear through the removal of the threat of violence from everyday life (Krause 2001, 13; for an overview of the concept see Security Dialogue 2004).

The development of the small arms agenda has continued apace since the late 1990s, and has been woven into the conflict–security–development nexus that has gained currency. As discussed further in Chapter Six, small arms are widely understood to be harmful to development, especially in 'those developing countries in which public institutions, such as police and health services, are predatory or failing' (Small Arms Survey 2003, 125). Small arms are thus widely understood as one element of the wider problem of (in)security in the South. This post-Cold War understanding is deemed to take account of the changed nature of international security in the aftermath of the Cold War, and pay greater attention to the effect on individuals and communities.

The progressive impetus to this shift notwithstanding, it has two significant effects that deserve scrutiny. First, the post-Cold War threat agenda, new wars argument and human security impulse tend to isolate the forms of violence in the South from the historical, political-economic and international contexts in which they occur. Analysis tends to start from the assumption that states should be internally pacified, with a monopoly on violence, outward-facing militaries and liberal domestic social relations. However, as discussed in Chapter Seven, this assumption is based on an idealised history of European state formation that is inadequate as a starting point for analysis of (in)security in the South. Second, it is notable that both the academic literature and NGO activity on the arms trade tend to separate the trade in small arms from the wider arms dynamic, to the exclusion of the high-technology end of the spectrum. Academics in the 1990s argued for small arms to be considered separately from the wider trade in conventional weaponry, based on the human security impacts of the changed international security agenda (Garcia 2006, 41). However, one effect of this is to isolate the small arms trade from the world military order of which it is a part.

These two criticisms point to the need for a single analytical frame for understanding organised violence. That is, the social relations of organised violence as perpetrated through the use of small arms need to be related to those perpetrated through major conventional and nuclear weaponry. There is a tendency in the academic literature and in political practice to focus on small arms as a particular category of weapons and on technical fixes to the problems of insecurity in the South. However, as Cramer argues, technological change 'has pushed war in contradictory directions: on the one hand it has made war cheaper and simpler and in a way more childish; on the other hand advanced industrial countries have been developing aerial bombardment, aiming to minimise civilian casualties, but at immense expense' (2007, 77). But while these different types of violence have different social logics and different effects, they 'must be studied together' (Cramer 2007, 72).

Post-Cold War arms control

Arms control has long been a staple of security studies. Dominant Cold War narratives of state security and military threats emphasised nuclear weapons and major conventional weaponry. Arms control initiatives – such as the Strategic Arms Reduction Treaties and the Nuclear Non-Proliferation Treaty – were state-led, dominated by diplomats, the military, and academic experts, framed in terms of military threats to state security, often highly bureaucratised and cumbersome in process, and giving rise to substantial verification commitments (Cooper 2006a, 368-70). In the renegotiation of the traditional security studies agenda at the end of the Cold War, arms control continued to refer to state-based, major conventional forces (Booth 1994; Chipman 1992) until the rise of the small arms agenda in the mid 1990s.

Notably, several of the more successful post-Cold War arms control regimes have related to particular categories of weapons, such as landmines, cluster munitions, and small arms (Cooper 2006a). The focus on such 'pariah' categories of weapons (Cooper 2000) at first sight seems progressive, in that it isolates particular technologies, facilitating speedier agreement and action, and changes weapons from an issues of state security into a humanitarian or development issue. However, it has two effects above and beyond this. First, it leaves unchallenged the wider patterns of military production and trade from which such technologies emerge; and second, it serves to reproduce the South as a site of intervention.

The pariah agenda, and in particular concerns around the scourge of small arms, has become almost synonymous with the problem of the arms trade because of the supposed transformation in the nature of conflict brought about by the end of superpower rivalry and the rise of the dark side of globalisation. Associated academic narratives of the new wars and human security have lent credence to the agenda, emphasising that the easy availability and use of small arms, their low cost and durability make them a particularly dangerous technology (Kaldor 1999; Muggah and Berman

2001; Liotta 2002). Alongside 'drug trafficking, international crime, and bloody civil wars raging around the globe' (Lumpe *et al.* 2000, 1; Boutros-Ghali 1995), small arms have become emblematic of the post-Cold War international security agenda.

There are thus two tendencies in arms control. First, supplier-based non-proliferation models of arms control, exemplified during the Cold War but persisting since its end, create a technical, apolitical image that divides states into suppliers and recipients, allowing the former to define what counts as destabilising accumulations of weaponry and to act to control from the supply side (Mutimer 2000). Commitment to non-proliferation keeps disarmament off the agenda, further entrenches the view that Northern preponderance is the key to international peace and stability, and focuses attention on weapons that are supposedly inhumane, uncivilised or pariah (Krause and Latham 1998; Cooper 2000). Second, a more holistic approach to arms control has emerged since the late 1990s that focuses both on addressing the demand for weapons, in particular small arms, as well as the motivations for their supply. Anti-proliferation efforts thus combine with a more interventionist transformatory demand-side approach, with the effect that the South is simultaneously disconnected from the North and the international system of which it is a part, as well as produced as a site of intervention.

The shift towards a more human security-oriented post-Cold War arms control agenda therefore has significant unintended effects. Arms control remains asymmetrical, aimed at particular categories of weapons or recipients and more generally legitimises disparities in military capability and military spending across the world and in particular further entrenches the predominance of the US within this. While NGOs have been pivotal in the successes of post-Cold War arms control, they are marginalised in terms of the 'hardness' of the issues they are allowed access on. That is, they have influence as long as issues can be framed in terms of human security or humanitarianism, but they remain (self-)excluded on harder matters of military security. And in relation to those issues on which they do exercise influence, the practices they participate in often serve to reproduce North–South hierarchies.

Two sides of the same coin: complementarity of imperial practices

The challenges of managing the arms trade and promoting arms control are conventionally understood in terms of the risks and opportunities of globalisation. The gains of positive globalisation (such as high-technology development and the jobs associated with it) need to be promoted, and the dark side of globalisation (e.g. weapons proliferation) protected against. Whether in its post-Cold War guise or more recent War on Terror incarnation, the globalisation agenda is dominated by issues such as international terrorism, the illicit drug trade, the spread of weapons of mass destruction, conventional weapons proliferation, failed states and so on.

In contrast to this understanding, military production and trade and arms control can fruitfully be understood as two sides of the same coin. They are the iron fists and velvet gloves of globalisation, both premised on a hierarchical relationship between North and South. They are both mechanisms of interconnection in international relations, articulated in different geographical areas and according to different logics of security. The North is assumed to stand outside of the problem (of weapons proliferation and/or conflict, insecurity and underdevelopment), and the South becomes a site of intervention, control or concern. The maintenance of state coercive capabilities at home and abroad and the promotion of human rights and good governance via the removal of small arms from Southern societies are two sides of the same coin. While not an instrumentalist project or functionalist imperative, the effects of these two sets of practices are oriented towards the promotion of stability in an internationalising capitalist state system and the prevention of autonomous or alternative development.

Both practices contribute to a post-Cold War shift towards the promotion of polyarchy – the formal trappings of liberal democracy, such as elections, without significant socio-economic change – as a preferred mode of engagement with the South. In this, '[a]s a general rule, authoritarian regimes are supported *until* or *unless* a polyarchic alternative is viable and in place' (Robinson 1996, 112-13, emphasis in original). Thus, while postwar US foreign policy revolved around the development of alliances with authoritarian and repressive regimes, today's polyarchic alternatives involve 'elite minority rule and socioeconomic inequalities alongside formal political freedom and elections involving universal suffrage' (Robinson 1996, 356). In particular, policies of 'democracy promotion' are 'aimed not only at mitigating the social and political tensions produced by elite-based and undemocratic status quos, but also at suppressing popular and mass aspirations for more thoroughgoing democratisation of social life' (Robinson 1996, 6).

Despite this recent trend, Robinson makes a key caveat. He argues that during the transitions from military to civilian rule that have occurred in many parts of the South, states' coercive apparatuses have remained intact, as they underpin the wider social order. Demilitarisation is thus 'controlled' and never total (Robinson 1996, 64-6). The promotion of polyarchy is 'a general guideline of post-Cold War foreign policy and not a universal prescription' (Robinson 1996, 112) – authoritarian regimes are left in place and supported where they are too strategically important to risk. The uneven promotion of polyarchy is exemplified through the arms trade, which provides the means of coercion that ultimately back up state regimes.

Human rights, good governance, and conflict prevention are corollaries of democracy promotion and key features of the polyarchy agenda, and speak to the integration of small arms work into wider programmes of conflict prevention, poverty reduction and defence diplomacy. So for example, small arms projects include the removal of weapons from societies in the South and demilitarising them more generally, as well as shoring up state capacities for coercion. Notably, these efforts at demilitarisation happen only in the South, but not everywhere in the South. They are focused on Sub-Saharan Africa and Eastern Europe, for example, but not the Middle East.

Controlling small arms in the borderlands of the South is oriented towards the containment of conflict and the '[transformation of] the dysfunctional and war-affected societies' of the South into 'cooperative, representative and, especially, stable entities' (Duffield 2001, 11). Elsewhere, where it would be too risky to attempt such measures, incorporation is preferred through the export of weaponry to states. Small arms clean-up programmes thus tend not to be carried out in states to which the major arms-producing states have a record of exporting weaponry, and support for authoritarian regimes continues, with initiatives like security sector reform not being attempted in states such as Saudi Arabia (see e.g. Laipson 2006). And at times when the institutionally more powerful elements of the state deem it necessary, hard security takes precedence over soft security, such as when funds from the UK's Global Conflict Prevention Pool were used to pay BAE Systems to maintain Tornados flying in Iraq (Hencke 2008).

Understanding the arms trade in terms of imperial relations is neither to reduce political violence in the South to US or Northern machinations, nor to insert the South into a pre-given structural position. Rather, it is to appreciate the historical context that shapes contemporary practices. 'Quasi-imperial' forms of rule certainly exist within the South (Shaw 2002), but they are often overlain by wider sets of North–South relations that provide significant context in which Southern actors pursue their own agendas.

Indonesia's occupation and repression of East Timor, Aceh and West Papua is a good example of quasi-imperial rule (Shaw 2002). But state formation and development in Indonesia have also been significantly shaped by US-led interventions to control its integration into the global capitalist economy, and by Dutch colonial rule prior to that. So Indonesian state agency must be read alongside the dynamics of European and US efforts to maintain asymmetrical relations.

Southern agency takes a variety of forms. In Saudi Arabia, for example, assistance from the European world has been central to the consolidation of state power, but this does not mean that Saudi leaders have not exercised agency (Bromley 1994; Gongora 1997; Nonneman 2001). On the contrary, they have played military suppliers off against each other in order to consolidate their rule, build up the institutional basis of the state, and create room for manoeuvre in the international system. India's postcolonial security imaginary has self-consciously been constructed as an alternative to superpower rivalry during the Cold War (Muppidi 1999, 128). Yet its nuclear tests of 1998 signalled, in the words of Arundhati Roy, capitulation to the 'ultimate coloniser' and the destruction of its noncolonial approach to world politics (quoted in Muppidi 2004, 99). Similarly, while India attempts to gain self-sufficiency in weapons production, thus resisting domination of supply by either the US, Russia or the UK, in doing so it further reproduces dominant narratives of statehood as requiring a modern, technology- and capital-intensive military.

Assessing NGO arguments

Having reconsidered the arms trade as an indicator of a world military order marked by imperial relations, how are we to assess the reformist and transformist arguments of the various NGOs? The first point of departure is their overall stance towards the arms trade. CAAT's abolitionist approach marks it out from the rest of the NGO community. As an organisation, CAAT does not espouse a pacifist position however, as it aims to be an umbrella organisation broad enough to cover a variety of political and ideological positions in relation to the trade. The other NGOs accept the legitimacy of the trade, often implicitly, by claiming not to take a position on the arms trade. However, such a stance *is* a position: a reformist one. The claim that CAAT engages in excessively 'political' action is a key means of sidelining it, along with its role in protest and direct action (discussed in Chapter Three).

Abolition of the international arms trade would obviously require more radical change beyond existing parameters than regulatory changes to ensure that all international arms transfers are legally authorised by states in line with internationally agreed standards. While an abolitionist agenda is widely dismissed as unrealistic, it does not deny the need for incremental steps along the way. And the interim steps demanded by CAAT – such as an end to subsidies, an end to exports to controversial recipients – are actually in line with demands of the other NGOs. This is not to argue that an abolitionist agenda is the only transgressive position. While none of the NGOs advocate this, a postcolonial position focused on resisting liberal or other forms of imperialism could argue in favour of arms transfers to state or non-state actors in the South. While such a position is not abolitionist, it is transformist in that it would change the parameters of the arms trade, aiming at a fundamental change in the world military order in which a greater proportion of actors had more equal access to the means of violence. Returning to agendas advocated by NGOs, a regulatory vision can be transgressive if it includes changes from within the existing order that could radically change the world military order. So, as discussed in Chapter Four, the work done by BASIC and Saferworld as well as CAAT on the economic subsidies on arms exports does not require an abolitionist framework to be transgressive, as a radical diminution of UK arms exports would have a significant impact on international relations. However, the regulatory approach of BASIC and that of the Control Arms NGOs can be differentiated according to their overall vision and wider argument, in that BASIC's regulation is transformist, while that of the others is reformist.

The NGOs also differ in their account of the source of the problem of controversial arms exports. For the Control Arms NGOs, the problem is largely one of weak control regimes, loopholes and poor decision-making, exacerbated by developments such as the weakening of export controls under the War on Terror. At the national and regional level, they focus on providing policy suggestions as to how weaknesses in the control regime can be remedied, such as extraterritorial controls on arms brokering, re-export clauses in licensed production agreements, robust end-use monitoring of arms exports, and a system of prior parliamentary scrutiny of arms exports. Further, their proposals for an Arms Trade Treaty rest on states' existing responsibilities under international law. However, if these responsibilities exist in international law, this begs the question why states do not already abide by them. Taking a supposedly apolitical approach is the only possible strategy when trying to generate reform via the UN, and is already difficult enough as it is, but the faith in regulatory mechanisms and legal reform obscures the impetus to the scale and nature of the arms trade, and the social forces that maintain it.

For CAAT, in contrast, controversial exports are often the result of deliberate government policy rather than an aberration. The impetus to such policies stems from the relationship between arms companies and the UK state, which creates a pro-export orientation towards licensing policy. While this approach focuses on an area left largely untouched by the other NGOs, it fails to address the existence of a complex bureaucratic licensing system that claims to regulate arms exports. And while the focus on the relations between the state and capital accounts for a significant element of the pro-export orientation, the wider geopolitical relations that are buttressed by arms sales are left untouched. That is, claims of national defence and state sovereignty, which carry significant ideological weight, cannot be reduced to the state–capital relation: militarism and capitalism should not be conflated. CAAT does not take a position on whether arms companies should be renationalised, but arms companies could be just as acquisitive if situated within the state (Dunne and Smith 1992, 109).

Relatedly, the other NGOs have a different view as to the major culprits of the international arms trade. There has been a shift within the Control Arms community: in the late 1990s and into the early 2000s, they focused on the UK as one of the world's major exporters, holding the government to account and assisting it in the implementation of a new control regime. They acted as watchdog and advisor to the government on the EU Code process, for example, as discussed in Chapter Three. With the launch of the Control Arms campaign in 2003, however, they shifted their focus towards the international stage, with an increased focus on getting states such as Russia, China and other growing non-European exporters on board. While the US has exercised opposition to the Arms Trade Treaty, NGOs are less concerned with getting it on board as its national control regime is already one of the strongest in the world, on paper.

The Control Arms campaign thus signals a shift away from improving European control practices to encouraging the rest of the world, particularly Russia, China and Southern producers, to improve their export controls. This is on the basis that such exporters often do not have human rights clauses written into their control regimes, if they have control regimes at all, and are more likely to transfer weapons to conflict zones. However, the shift of focus away from European producers is problematic for two reasons. One is the scale of the arms trade: Southern producers will never accept restrictions on their involvement in the trade if the scale of exports from or military production within the European world does not diminish. And while European states do have human rights concerns explicitly worked into their control regimes, they also contravene these if needs be, using arms transfers as a deliberate element of foreign policy.

For CAAT in contrast, the focus remains the UK state, because the UK is a major arms exporter and the campaign is a nationally based coalition that aims to hold its own state to account. BASIC, meanwhile, focuses on the transatlantic relationship, pushing for a fundamental change within Western security thinking. One argument against CAAT's national focus is that of globalisation: the internationalisation of military production makes it insufficient to focus on a single nation state. In this sense, the Control Arms NGOs are better at addressing the internationalisation of the industry and of state policy, and the transformation of state power. However, CAAT is right that globalisation is not simply undermining the state: the restructuring of the arms industry is facilitated and regulated by ongoing high levels of state intervention. While ownership and production are becoming more international, companies are not losing their home base. So these are processes of internationalisation rather than transnationalisation, predominantly. The arms industry remains largely exempt from the liberalisation of trade rules seen in many other industries, so military production is less internationalised and more dependent on the state than other forms of production. Governments try to articulate processes of internationalisation as being in the national interest; states are being transformed rather than necessarily weakened.

Regardless of their overall orientation, a key point of agreement amongst NGOs is that the main problem with the arms trade revolves around exports, in particular to the South. For the Control Arms NGOs, 'irresponsible' exports are illicit or unregulated transfers to the South, by either Northern or Southern suppliers. While it is true that two-thirds of the *international* arms trade flows from North to South, trade within as well as between Northern producers is also significant. Northern states are the world's largest military producers, with the US dwarfing even its nearest competitors. By taking exports as their starting point, thereby equating the arms trade with international arms exports, NGOs disconnect international transfers from wider patterns of military production, which includes military production within the North as well as North–South transfers.

The partial exception to this is CAAT, which allies its call for the end to the international arms trade with the goal of progressive demilitarisation within arms-producing countries. Its focus on the relationship between arms capital and the state within the UK and its ongoing work on the economic arguments around the arms trade (discussed in Chapter Four) thus speak to both its goals. For CAAT, severing the close relationship between arms capital and the state would help orientate the UK economy towards civil production more fully as well as end exports to oppressive regimes and to countries involved in armed conflict, in regions of tension, or in which social welfare is threatened by military spending (CAAT, no date, a). In practice, however, when criticising the arms trade, CAAT focuses on the international trade, in particular controversial exports to the South, albeit with a different objective and argument compared to the Control Arms NGOs. Its conception of the trade includes that between arms companies and the UK government, but domestic procurement or military posture rarely feature in its everyday campaigning. As such, a problem with this practical manifestation of an abolition agenda is that it runs the risk of freezing an unequal status quo by entrenching military disparities between producers and non-producers if it fails to challenge Northern military budgets and domestic procurement. Arguing for an end to controversial exports – which are usually exports to the South – without a wider argument about the structure of military power in international relations leaves the military dominance of Northern states intact and does nothing to challenge hierarchical North–South relations.

NGOs' focus on the South, although understood as a progressive shift, serves to disconnect it from the wider structures of the world military order and thus reproduce it as a site of intervention. NGOs concentrate on the South itself rather than the North–South relationships fostered through the arms trade. NGO efforts to tackle small arms proliferation accept the parameters of the international arms trade, that is, the military disparities between the North, in particular the US, and the South, as these are not deemed to be part of the problem. The problem is defined as conflict in the South; the solution is better regulation of the arms trade, in particular that part of the arms trade that most affects the South, that is, the illicit trade in small arms. But this assumes conflict in the South to be unconnected to wider disparities and inequalities in international relations, of which military disparities are one indicator. This is problematic both in terms of tackling the problems posed by the arms trade but also in terms of the representational politics of North–South relations – it is another way in which imperial relations are reproduced.

Arms control has traditionally been marked by a North–South distinction, in that traditional supply-side models rest on the assumption of Northern benevolence and military preponderance, and serve to freeze an asymmetrical status quo. The post-Cold War shift to small arms control, with a focus on conflict, insecurity and underdevelopment, similarly focuses on the South as a site of intervention, disconnected from the wider structures of international relations of which it is a part. The successes of particular arms control campaigns, such as those on landmines and cluster munitions, should not blind us to the fact that the focus on particular categories of weapons and their framing in humanitarian terms does not challenge the wider patterns of military production and trade of which they are a part and, if anything, serves further to entrench the South as a site of Northern benevolence.

Conclusion

The arms trade and its control are a key means of interconnection in international relations. It is not just hierarchical but imperial, reflective and generative of wider practices of coercion and rule in international relations. While military production and trade provide the iron fists of empire, small arms control provides the velvet gloves. The NGO community articulates two broad sets of arguments about the arms trade, one reformist and one transformist. This chapter has explored the similarities and differences between them, arguing that what they share is that they largely ignore Northern military production, severing international exports from the wider arms dynamic. When NGOs talk about the South, they disconnect it from the world military order and reproduce it as a site of intervention. The idea of the world military order allows us to highlight the links between the trends in the arms trade: intra-Northern production and trade, North–South transfers and small arms proliferation. These three trends in the arms trade are discussed in chapters Four, Five and Six respectively. The next chapter focuses on the broad contours of NGO strategy, as argument and strategy are mutually constitutive.

Note

1 'Small arms and light weapons' refers to military-style weapons and commercial firearms that can be carried by either a single person or several people serving as a crew and is a term based on the 1997 UN *Report of the Panel of Governmental Experts on Small Arms* (Small Arms Survey 2001, 8). Here, the term 'small arms' is used to cover both categories unless specified. See also Small Arms Survey (2009, 8–11) for a discussion of the various issues around defining this category of weaponry.

3 • NGO Strategies and the Disciplining of Global Civil Society

What strategies are most appropriate to achieve the NGOs' various goals? Is it more effective to provide information to elite actors and try to promote incremental reform, or to take a more confrontational stance and try to embarrass the government into making policy changes, or can both approaches be combined? The basic trade-off is this: a reformist argument that remains within the parameters of the government's and arms companies' frameworks is more likely to get taken up by those actors, and incremental changes around the edges may result. Yet such an approach has no answer to the question of why controversial exports continue, let alone why the scale of the arms trade continues unabated. More transformist approaches, in contrast, stand outside of the mainstream frame of reference and have a more transgressive and thus counter-hegemonic vision. However, the likelihood of such changes being made decreases, the further outside of the mainstream one moves, and the question remains as to whether the social forces exist for such a vision to take hold.

The outsider position is widely seen as a useful foil for the insiders, in that it creates political pressure that allows insiders to make progress on incremental reforms. However, the reverse is rarely considered: that insiders co-opt the available room for manoeuvre, thus further diminishing the possibility of more fundamental change. In addition to this question of the cumulative impact of NGO activity, the capitalist structuring of civil society also requires us to reconsider questions of NGO effectiveness. Rather than conceiving civil society as substantively separate from the state and market, we can think in terms of dual networks: one comprising elements of the state allied with arms capital, and one consisting of other elements of the state and NGOs. Assessing NGO activism against the backdrop of such a constellation of social forces demonstrates the challenge that faces them.

Insiders and outsiders: NGO strategies

NGO objectives and arguments are shaped by and contribute to the strategies they use to try to effect change. Generally, the closer an NGO's argument is to the received understanding of an issue, the more insider its strategy; the further away it is, the more outsider. Reformist arguments tend to lend themselves to an insider strategy, in which groups attempt to build a consultative relationship with government on policy matters (Grant 1978). Tranformist arguments tend to lead to an outsider strategy, in which groups are less able or willing to build a consultative relationship with government (Grant 1978). Of course, these differences are a heuristic device and constitute a spectrum. NGOs often use a mix of strategies or at least tactics, but understanding NGOs' overall orientation towards an issue remains useful.

NGOs using an insider strategy are better able to generate change, but that change will likely be incremental, technical in nature and featuring high potential for co-option. Outsider strategies involve more transformatory demands that cannot be accommodated in the current state of affairs, but are less likely to be listened to by policymakers. Broadly speaking, IANSA, International Alert and Saferworld couple an insider strategy with a reformist vision; Amnesty and Oxfam combine insider and outsider strategies with a reformist understanding; CAAT adopts an outsider strategy allied to a transformist vision; and BASIC combines a transformist vision with an insider strategy. In addition, coalition work through the UK Working Group on Arms and the Control Arms campaign is a deliberate attempt to generate cumulative impact through collaboration, precisely because of the different mandates and strategies that each organisation has. CAAT is not involved in either of these fora, partly through exclusion by the other NGOs and partly through self-isolation.

Arguments about insider and outsider strategies are typically employed in a pluralist framework of analysis. That is, while actors do not all have equal access or influence and are not all equally well-endowed in terms of resources, analysis proceeds on the basis that 'power in society is fragmented and dispersed' as well as non-cumulative (Grant 1989, 26-7). This means that government 'is not identified with any particular interest but rather acts as an independent arbiter between interests' (Marsh 1983, 10), such that policy is made through a 'competition of viewpoints' (Jordan and Richardson 1987, 46). However, the capitalist context of civil society and, as will be discussed in more detail below, the relationship between the state and arms capital mean that we need to re-read NGO activity in a

different light. Some of the NGOs under analysis here have been effectively integrated into the UK state through their work with the Department
for International Development (DfID). Moreover, the UK government
does not act as an independent arbiter between interests as arms capital
is structurally privileged. NGOs are thus in a disadvantaged position
compared to arms companies and lobby groups. Arms capital is the ultimate
insider group as it is integrated into the state and exercises disproportionate
influence compared to even the most insider NGOs, which are allied to
weaker fractions of the state.

Amongst the NGOs, an insider strategy is best exemplified by Saferworld, which works 'with governments, international organisations and
civil society' for the prevention and reduction of violent conflict and
promotion of cooperative approaches to security (Saferworld no date,
c). It has a strategy of putting itself in policymakers' shoes and making
constructive proposals, according to one member of staff, which wins it
the respect of policymakers and thereby access, allowing it to see behind-
the-scenes debates and find points of leverage it would not otherwise be
able to identify. With such an approach, the distance they are able to shift
government is less, but the prospects for shifting them at all are greater,
according to another staff member. As described by another staffer, Saferworld's main engagement with government on export control issues is with
DfID and the Foreign and Commonwealth Office (FCO), with whom it
has a 'very good' relationship; it has a 'reasonably good' relationship with
the Department for Trade and Industry (DTI) and Ministry of Defence
(MoD) but is 'kept at arm's length' on licensing issues. It therefore backs
up its advocacy with media work, as a dual strategy of simultaneously
conducting confidential work with government and maintaining a climate
for change, according to one senior member of staff.

Saferworld's insider strategy is particularly pronounced in its small arms
work in the South. Its engagement in mapping exercises to document the
extent of small arms proliferation, the development of regional control agreements, the provision of advice to DfID and Southern governments, and its
capacity-building work with local partners, are all undertaken through a
strategy of partnership with DfID and with elements of the state and civil
society in the South. The epitome of this insider strategy can be seen in the
award in 2002 of an MBE to its then director for services to the prevention of
armed conflicts and his inclusion in 2006 on the UK government delegation
to the UN Review Conference of the 2001 Programme of Action on Small
Arms. Saferworld accounts for the lion's share of government funding for small

arms work. Between 2001 and 2007 DfID, which is the biggest institutional funder of the UK-based NGO sector as a whole (see Wallace 2003), disbursed £11,542,677 from the Global Conflict Prevention Pool (pooled DfID, FCO and MoD resources for work that includes small arms issues, disbursed by the former) to UK-based NGOs for small arms projects, of which £10,214,677 was to Saferworld; the rest went to International Alert and Oxfam (DfID 2007a). Through its partnership with the state in its small arms work and its credibility with policymakers that means its suggestions on arms export licensing are taken seriously, often finding their way into government policy, Saferworld has to a considerable degree been integrated into the UK state.

International Alert's work on small arms and peacebuilding issues also proceeds via an insider strategy. It engages in capacity-building, mediation and dialogue and conducts policy analysis and advocacy at national, regional and international levels. More specifically, it engages in activities such as monitoring the implementation of arms proliferation agreements, making national and international policy recommendations on security issues, facilitating discussion of security issues between government and civil society (defined as NGOs, journalists, academics), and training civil society actors in security sector reform (International Alert 2006b). This is a similar strategy to Saferworld's small arms work; indeed, the two organisations collaborate on the Biting the Bullet project, launched in 1999 in conjunction with the University of Bradford to contribute to official small arms debates. What the two NGOs' approaches share is an emphasis on expertise, credibility and constructive policy proposals: they are thus emblematic of an insider strategy.

IANSA, also focused on small arms, seeks to raise awareness with policymakers, the media and the public as to the threats posed by small arms to human security and human rights. It promotes policy development, public education and research by civil society actors, fosters collaboration between civil society organisations and facilitates their participation in global and regional processes; it also seeks to promote survivors' voices. It claims to represent civil society on the international stage and summarises its strategy as policy advocacy, communications and civil society engagement (IANSA no date, a).

BASIC also enacts an insider strategy, engaging in what a senior staffer describes as 'insider advocacy', conducting research in order to engage in information-sharing and public education. It tracks weapons sales, conducts research and analysis of 'future decision points in transatlantic security policy', assists in 'the development of transatlantic security policies, policy making and the assessment of policy priorities' and promotes public understanding, informed debate and creative solutions through 'decoding complex defence and security material' (BASIC no date, b). It engages in advocacy in partnership with similar organisations in order to 'pressure governments to establish effective control and monitoring of conventional arms' (BASIC no date, a). These activities contribute to its 'unique role as a transatlantic bridge for policy makers and opinion shapers' (BASIC no date, b). This is a similar approach to Saferworld on arms transfer issues: the provision of expertise to elite actors (such as government officials) backed up by media work and dialogue with other civil society actors.

Unlike Saferworld, however, and although it was a co-founder of the Biting the Bullet project, BASIC moved away from its intensive work on small arms issues in the late 1990s to focus more on nuclear, biological and other WMD (weapons of mass destruction) arms control and disarmament, as well as security strategies such as peacekeeping, conflict management and prevention, and the control of terrorism. And since 2007 it has focused predominantly on global nuclear disarmament. As well as a shift in focus, this has led to a divergence in strategy, with BASIC much less integrated into the state on arms control issues than Saferworld. According to a former staff member, at its inception BASIC saw its role as translating the radical demands of CND and CAAT into something with which government officials could cope, so its insider role was predicated on more radical groups being active as well. This orientation is ongoing within BASIC: a combination of elite-level advocacy, often through personal contacts, is buttressed by the outsider work of groups such as CAAT, according to one staff member. This stems from a belief that arms exports are a political battle, not an intellectual one. This is a combination of insider and outsider strategies: insider work is undertaken in an attempt to create space for more transformist understandings of the problem posed by the arms trade, and outsider work is done in an attempt to ensure the most transgressive solution is adopted.

Amnesty International and Oxfam also attempt to combine insider with outsider strategies at the institutional level: they engage in both mass, popular campaigning and insider advocacy. However, as discussed in Chapter Two, they have a more reformist argument concerning the arms trade than does BASIC. Amnesty International believes that the most effective approaches to government take place 'in an environment where it is possible to establish positive long-term relationships with individuals and institutions, even where major disagreements persist.... AI must be seen as a respected and credible organisation' (Amnesty International 2001, 265).

The emphasis on respect and credibility is typical of an insider strategy, and Amnesty's capacity to mobilise public pressure is understood as a key element of its credibility as a lobbying organisation.

Amnesty's approach can be described as the provision of constructive criticism. As a UK section staffer described it, the combination of a credible policy solution and mass public pressure is needed to give government a reason to listen to them, in contrast to an outsider approach that just provides criticism. One way in which Amnesty attempts this is through the use of existing international law as a base for arguments and interventions it thinks governments would respond to. As an International Secretariat researcher put it: 'We start with existing law, and governments' existing obligations, and ask why aren't you doing these? Or we say that these standards aren't good enough.' The use of existing law is thus also part of a wider standard-setting strategy, which has always been backed up with public pressure. The first Secretary General of Amnesty International, Martin Ennals, argued that 'The only power which an organisation such as AI can hope to exercise is that of publicity or the threat of publicity;' hence the importance of mass public opinion (Ennals 1982, 79). This was echoed by a UK section campaigner, who argued that public opinion can help push issues up the agenda. Central to both research and campaigning is the emphasis on impartial and credible research findings.

Amnesty's approach to the arms trade is part of its wider mission to 'campaign for internationally recognised human rights to be respected and protected for everyone', independently of governments, political ideology, economic interest or religion (Amnesty International no date). Its reputation for impartiality and non-political activity, taking 'no stand on political questions', has been central to its development and success (Clark 2001, 12). As Stephen Hopgood argues, Amnesty's authority is moral in nature and this rests on detachment and the characterisation of human rights as non-political, characterised by an International Secretariat staffer as 'not grinding political axes' but rather 'providing the information that others could grind political axes with if they wanted to' (Hopgood 2006, 14). In this sense, there is something of a synergy between Amnesty and CAAT: while the key point of difference between them is whether they take a 'political' or an 'apolitical' stance, Amnesty's emphasis on 'conscience (rather than simply compassion)' and non-violence (Hopgood 2006, 65) resonates with CAAT's approach to campaigning.

However, there is a tension within Amnesty, between 'keepers of the flame' and 'reformers'. The former are 'the guardians of the Amnesty ethos' and proponents of moral authority, taking the position of witness; the latter are 'more engaged, more of a movement', tapping into 'demonstrative protest, political solidarity, and social change' and acting as advocates (Hopgood 2006, 11-14). Work on arms issues, a thematic issue that cuts across country programmes, exemplifies this tension within Amnesty's work. On the one hand, it is symbolic of the rise of reformers, who want a more political approach to human rights, moving beyond traditional political and civil rights, deploying truth in pursuit of social change, and challenging the 'Work On Own Country' (WOOC) rule (through the central role of the UK section researching and campaigning on UK arms export control issues). On the other hand, Amnesty claims to take no position on the arms trade, and focuses on the suffering caused by the use of arms in human rights violations and abuses – in this sense, its work on arms is just like its wider work on human rights issues.

The WOOC rule is itself an example of the tension in Amnesty's overall strategy. The rule prohibits members from researching or campaigning on human rights issues in the state of which they are a citizen, and is seen as an expression of Amnesty's fundamental principles of impartiality, independence and international solidarity, which form the basis of its credibility (Amnesty International, quoted in Hopgood 2006, 98). However, the practical manifestation of the rule has been to make it difficult to generate membership outside of the West, and those from the South who do join tend not to be grassroots activists. The rule reinforced 'a culture of expertise that delegitimised those whose expertise came from experience (the expertise of those most active in human rights struggles in their own countries)' (Hopgood 2006, 98-100). It has been one of the reasons why Amnesty has been 'structurally unable to build a constituency' outside of the West (Hopgood 2006, 174).

Like Amnesty, Oxfam combines research with campaigning. However, Oxfam engages in advocacy more than bearing witness: its work has a more practical policy impulse than Amnesty's. It aims to maintain pressure to create the political will so that the most ethical judgement is always made, according to one former policy adviser. This is aided by the emergence of arms export scandals, which are good for media coverage and enrage local campaigners, in the view of both a policy adviser and a campaigner. Advocacy and campaigning are mutually reinforcing: for policy advisers, while the campaigning aspect of Oxfam's work is complemented by its advocacy work, advocacy also needs campaign pressure to be successful. And according to one of its campaigners, through its size and reputation as a household name, it is able to engage in both successfully. A former policy adviser articulated one of the organisation's claims to success as having converted people at ministerial level and ensuring they internalise the Oxfam message.

Oxfam's greater policy impulse and reliance on state funding compared to Amnesty, sometimes mute its ability to be critical of governments. While there is a broad similarity between Oxfam and Amnesty in terms of strategy, in their combination of insider advocacy with mass campaigning, as well as a wider shared understanding of the source of the problems of the arms trade, there is a different attitude towards government. For example, an Oxfam campaigner admitted that Amnesty can give 'a harder push' than Oxfam at times. A key issue for Oxfam is the balance between its work on arms issues and its other country programmes. There are some countries that Oxfam cannot discuss in relation to arms control in general or UK exports in particular because there is a risk that their employees' security would be endangered or that its humanitarian programmes would be ejected from the country, according to campaigners. Yet, in the eyes of Amnesty campaigners, these countries are also those that best illustrate the problems of the arms trade, such as Indonesia, Israel, Sudan or Zimbabwe. An Oxfam campaigner explained that the organisation also faces difficulties around attempting to remain politically neutral in conflict zones whilst also commenting on arms issues, for example in the Democratic Republic of Congo, Aceh and Colombia. As a humanitarian agency with a presence on the ground, Oxfam is supposed to be politically impartial. But decisions on arms sales are a political judgement, and that creates a conflict. This demonstrates the flipside to a strategy that combines insider and outsider approaches: causing offence to governments through more confrontational strategies on arms export control can damage credibility and put field programmes at risk.

The campaigning elements of Amnesty's and Oxfam's work give them something in common with CAAT. Whilst all three engage in campaigning, there is a fundamental difference between them, however, in terms of their attitude towards government and what they perceive to be the problem with the arms trade. Oxfam and Amnesty use their public campaigning as a means of backing up their advocacy message, which is based on an understanding that more effective policies and implementation would solve the problems posed by the arms trade. CAAT, in contrast, places more emphasis on public education, and does not engage in advocacy with officials as its primary course of action because 'If it's not a logical situation…it doesn't matter how many civil servants you talk to, because it's a political decision taken higher up', according to a CAAT staff member. In this view, the best argument does not always become policy, especially when there are vested interests involved.

As a result, CAAT deems the most appropriate avenue for action to be the democratic process, to use public pressure and the media to try to force the government to change, as one staffer put it. CAAT's public education and awareness raising work is a means to build public opposition to the arms trade, which backs up its policy engagement with MPs and officials. One former staff member described it as creating educated citizens in contrast to the 'consumers of campaigns' that groups such as Amnesty and Oxfam foster. CAAT believes public opinion 'has really changed over the last twenty years and is very anti-arms now', in the words of one staffer, and believes that only public opinion can force the government to change its ways. MPs and civil servants need a political reason to change policy: that is why public pressure is needed and arguments alone will not suffice, as one researcher described it.

As well as media and parliamentary work, CAAT runs thematic campaigns focusing on different aspects of the arms trade, and has historically run an annual Clean Investment campaign (aimed at encouraging local authorities, charities, religious organisations, health organisations and universities to disinvest from arms companies). An important part of its campaigning is non-violent direct action. Just as the clearest line of distinction between NGO strategies is CAAT's opposition to the arms trade as a whole, one of the means of distinguishing them in strategy terms is its protest activity. Not only is protest part of a more confrontational stance, it also makes CAAT the only organisation to direct pressure towards companies as well as the government. This strategy challenges the supposed separation between the economic and political realms: CAAT understands the problem to be one of state–industry relations, not just of government policy.

The strategies of Amnesty International, BASIC, IANSA, International Alert, Oxfam and Saferworld share a focus on credibility, a relationship with government and the provision of expertise. This means that even on issues where the NGOs do not agree with the government – as in the case of particular, controversial exports – they are still taken seriously. Opposing the government on specific issues and using mass campaigning (as Amnesty and Oxfam do) thus do not necessarily mean that a group is an outsider. Indeed, Amnesty International and Oxfam believe that the ability to mobilise public opinion is a key means of legitimation and credibility, and backs up their advocacy message. Insider activity requires a 'strategy of responsibility' (Grant 1978, 6), which requires that an organisation engages with its target and is seen as responsible and reasonable. While access to the machinery of government is an important part of an insider strategy, it generates constraints as groups have to show 'tact and discretion' in their dealings with government (Richardson, quoted in Grant 1978, 3). In addition, an insider strategy carries the risk of 'benign neglect' in which the government praises the aims of the group but does little to fulfil those aims (Grant 1978, 2). Insider strategies thus run the risk of co-option through the assimilation of challenge.

CAAT, in contrast, does not believe that the government can be persuaded by NGOs to change, because of the vested interests at stake. For CAAT, 'Presently, the only meaningful constraint on arms exports is political embarrassment' (CAAT 2006a). Its strategy is therefore more confrontational and is aimed at broad public education and mobilisation against the arms trade, rather than at persuading policymakers, who it believes will not listen to a message that threatens those interests. The biggest difference between CAAT and the other NGOs is that CAAT's strategy revolves around the argument that fundamental change is necessary and will not occur by improving the technical processes of export control. So, as described by a senior member of staff, while CAAT makes submissions to parliamentary and government consultations and meets with MPs and civil servants, it does not go through the export control regulations line by line to ensure they are as tough as they can be. The other NGOs and the government tend to see CAAT as making a political point and failing to act constructively. CAAT acknowledges this to an extent: as one staff member articulated: 'because we are anti the arms trade *per se*, it is difficult to say "reform it"'. CAAT acknowledges that there are practical steps to be taken to end the arms trade but believes the primary issue to be that of political will. CAAT therefore needs to act as 'a mosquito biting an elephant', according to one researcher. In this view, 'the government is a law unto itself so we need to be as scandal-mongering as possible'.

NGO funding

The main funding sources for NGOs are governments, foundations, charitable trusts, individuals and trading (although only Amnesty and Oxfam raise a significant proportion of their income through trading). Major governmental funders include the UK, Sweden, the Netherlands and Canada, as well as the EU. Amnesty International, however, does not accept money from governments for its work investigating and campaigning against human rights violations, although it does occasionally seek money from governments for other work as long as the funds 'are free to be used without compromising our aims and principles' (Amnesty International UK, no date, a). Funding is only accepted from what one International Secretariat staffer described as 'non-tainted' governments for human rights education materials and would not be used for items such as staff salaries and overheads. Amnesty prides itself on being independent and impartial and does not want to be compromised by its funding: as an IS staffer put it, 'it's not a consultant to the state'. And CAAT does not seek or accept government funding, UK or other, on principle.

In refusing government funding, Amnesty and CAAT make a clear statement of independence and do not have to make the trade-offs associated with criticising a funder. While the other NGOs maintain that they remain independent and impartial, they have to contend with the inherent disciplining of funding in additional ways. NGOs have to persuade organisations or individuals to fund them, and thus seek to appeal to their goals, values and interests. The predominance of more mainstream funders is often viewed in critical academic circles to mean that NGOs 'are inevitably drawn into supporting and even spreading many aspects of the dominant global agenda', becoming 'carriers of these concepts, values and practices' (Wallace 2003, 203). However, there is a wider integration of NGOs beyond funding and direct financial control, in that they are often active participants in the shaping of these concepts, values and practices, rather than simply carriers of them. This is most clearly demonstrated on the issue of small arms, on which NGOs have to a considerable extent led the way in developing international understandings and practices of small arms control.

Tensions associated with significant levels of DfID funding are the trade-off between programmes and the strategic rationale behind taking money from one of the world's largest arms-exporting states to work for tougher controls on the arms trade. For example, DfID funds over half of Saferworld's work, predominantly on small arms but also some of its work on the Arms Trade Treaty. One effect of accepting DfID funding for small arms work is that NGOs are more reticent in speaking out on issues of national export controls. Saferworld, for example, can either 'slam the government 100% and get nothing, or slam them 70% and get £2m from DfID, which lets you do your work in Eastern and Southern

Africa', as one senior member of staff described. If Saferworld did not rely on DfID money for fieldwork, it could 'hammer the government a bit more', according to a senior staff member. With the advent of the Control Arms campaign, however, Saferworld has moved away from criticising the UK government so much, and DfID even funds some of its research into the Arms Trade Treaty, indicating both shared policy assumptions and a collaborative relationship.

In addition to governments, foundations and charitable trusts also provide significant funding for NGO activity on arms issues. Key sources for arms trade NGOs include the Joseph Rowntree Charitable Trust (JRCT), the Polden-Puckham Charitable Foundation, the Ploughshares Fund, the Ford Foundation, the MacArthur Foundation, the Diana Princess of Wales Memorial Fund, Comic Relief, the Community Fund, and the Network for Social Change philanthropic organisation. While disciplining is inherent to funding, there are differences between foundations. For example, foundations such as Ford, Rockefeller and MacArthur can be broadly understood to engage in ameliorative practices associated with maintenance of the capitalist system (Arnove and Pinede 2007). In relation to arms control, this means a focus on ironing out controversial exports and small arms control, rather than a stance such as CAAT's which focuses on the state–capital relationship.

Under the War on Terror, large US foundations that have traditionally funded several of the UK-based NGOs, such as Ford, Rockefeller and MacArthur, have moved away from national arms export controls and towards anti-terrorism efforts such as the prevention of nuclear terrorism. The Ford Foundation, for example, which had funded Saferworld to the tune of £60,000 a year previously, will now only fund arms work that focuses on nuclear issues, and the MacArthur Foundation has moved away totally from funding export controls projects, according to a Saferworld staff member. JRCT, on the other hand, is a Quaker organisation whose grants are designed to support work to 'advance the cause of peace and nonviolence', such as organisations or individuals working on 'control *or elimination* of specific forms of warfare and the arms trade' (JRCT 2007, emphasis added), and the Polden-Puckham Charitable Foundation supports work to resolve conflicts and *remove the causes* of conflict (Polden-Puckham no date, emphasis added). Whilst these are not explicitly anti-capitalist foundations, they are anti-militarist. There are thus different priorities in play in terms of funding, with some more likely to be counter-hegemonic than others.

NGOs accept funding from a variety of sources, however, including both types of foundation discussed above. This suggests that there is no straightforward 'capture' of NGOs by funders' preferences. NGOs are able to pitch their proposals in a manner that speaks to the interests of the various funders. The fashions or trends in what funders are prepared to devote resources to remain significant, however. The big US foundations had more resources at their disposal, and their move away from national export control work poses a significant challenge to NGOs working in this area.

Individual or membership donations are also significant for Amnesty, CAAT and Oxfam. Nearly half of Amnesty International UK's income in 2004–5 came from membership contributions and subscriptions, and over 25 per cent came from appeals and donations (Amnesty International UK, 2005a). The majority of CAAT's income (approximately 80 per cent) comes from individual donations, with a smaller proportion coming from contributions from groups; nearly all of CAAT's active campaign work is funded by individual donations (CAAT no date, c). Over half of Oxfam's income in 2003–4 came from donations (Oxfam no date). Whilst membership subscriptions and donations from the general public generally count as unrestricted funding, which means they can be used to fund any aspect of work, gifts, donations and grants from major donors such as governments, foundations and trusts are likely to be restricted funding, meaning the money has to be used for a particular purpose.

Restricted funding is typically precarious, depending on funders' preferences and the wider financial and international climates. The impact of the dotcom bubble bursting in 2000, the War on Terror and the current financial crisis, all serve to restrict the availability of funds for NGO work on particular issues and signal the disciplining of NGO work through the wielding of financial power. More generally, the availability of funding reflects the issues that are in fashion with grant-makers at any given time, and through the increase in activity on such issues as a result of the disbursement of funding, the fashionability of certain topics is reinforced.

Amongst both government and foundation funders, small arms and conflict prevention in the South are popular post-Cold War and post-9/11 issues. These are thus relatively easy issues for which to get funding, markedly more so than national export controls. However, 'donors won't fund the same thing forever', requiring NGOs to be innovative and 'move forward' in the types of activity they carry out, as a Saferworld staff member put it. In the Saferworld case, for example, small arms projects have metamorphosed into community safety and community policing projects. In contrast, its national arms export control work is harder to fund. This has implications for organisational survival: according to a BASIC staff member, Saferworld has been able not only to survive but also to grow as a result of its funding from DfID, while BASIC and other NGOs are struggling.

Persuasion versus protest: NGO campaigns since the 1990s

The differences in strategy within the NGO world can be seen in a brief history of their campaigns since the 1990s. NGO activity took off with the campaign for an EU Code of Conduct to regulate arms exports from European states. Starting in the mid 1990s, NGOs worked behind the scenes and in the public realm to harness post-Cold War energies and capitalise on the election of a Labour government in the UK. This period also accentuated the differences between CAAT and the other NGOs and was the start of the divergence between them. Since then, NGOs have gone one of three routes: the Control Arms NGOs have led the way in calling for an international, legally binding Arms Trade Treaty via the UN, to regulate the trade in conventional weapons. CAAT has focused more intensely on the relationship between arms companies and the UK state, orchestrating a campaign to reduce the influence of arms companies on government policy. BASIC, meanwhile, has returned to its original mandate of working for transatlantic nuclear security, launching a 'Getting to Zero' campaign to maximise energy for nuclear disarmament. The different arguments and strategies put in play by various NGOs raise the question of cumulative impact: what is the overall effect of these different types of NGO activity?

EU Code of Conduct

NGOs' role in the establishment of the EU Code of Conduct, which harmonises export control regulations between EU member states, is a key moment in the history of NGO work on the arms trade. NGOs had a great deal of influence in setting the agenda of the incoming UK government, according to one senior civil servant, and the Labour Party, in its 1997 election manifesto, publicly committed itself to refusing arms sales to destinations where they might be used for internal repression or international aggression and to supporting an EU code of conduct on arms sales, while maintaining its support for a strong UK defence industry and defence through NATO. Once Labour was in government, its 'ethical dimension' to foreign policy specifically included concerted action to control the international arms trade (Cook 1997a, 1997b). The Labour victory kick-started the process of the emergence of national and EU-wide guidelines governing arms transfers.

Key NGOs involved at this early stage were Amnesty, BASIC, CAAT, Saferworld, and the World Development Movement.

The Labour government announced a new set of national guidelines on arms exports, the basic lines for which were defined in close collaboration with NGOs, according to a senior civil servant involved in the negotiations. The guidelines took on board 'several, but certainly not all, of the recommendations by British nongovernmental advocates' (Anders 2005, 187). According to John Kampfner, however, 'John Holmes, Blair's principal private secretary and top civil servant, went through it with officials from [foreign secretary Robin] Cook's private office for four hours, line by line, telling them to tone down various areas' (quoted in Phythian 2000a, 291). When the UK took on the EU Presidency in January 1998, progress began to be made on the development of EU-wide controls on arms exports, encouraged by NGOs. The UK's national guidelines became 'the basis for a British-Franco EU initiative to work toward a strengthening of common arms export controls' (Anders 2005, 182).

NGO strategies around the Code ranged from awareness-raising and helping shape governmental policy agendas through research and advocacy, to participating in the design of the Code and its eight criteria, and pushing for its implementation. Once the proposal was on the table at the EU level, NGOs engaged in a dual strategy of continued insider support through the provision of expertise, combined with a more vocal, critical approach – exposing weaknesses of the proposed Code via the media, for example – that tried to ensure the strongest possible wording of the Code, according to a senior Saferworld staff member. Within several months of the launch of the EU Code initiative, 'a network of over 600 European NGOs had emerged that endorsed the campaign' (Anders 2005, 185), albeit led by UK organisations in terms of technical expertise and strategic direction. According to a CAAT staffer, several of the European anti-arms-trade groups felt that the British groups were dominating the process, however, writing their own Code and only then asking others across Europe to come on board.

With the addition of Oxfam and Save the Children, the group of UK-based NGOs that came together to discuss and work on the EU Code metamorphosed into the UK Working Group on Arms, a coalition of NGOs working on arms control. The inclusion of Oxfam and Save the Children was described by a Saferworld staffer as broadening the spectrum of voices talking about arms control; the presence of large organisations within the coalition meant they could 'mobilise constituencies to which the British NGOs that originally proposed the code had no access' (Anders 2005, 185).

CAAT was involved in the early stages of the Code work but withdrew as it felt it could not endorse it, although it did not campaign against it (CAAT 1996). CAAT was opposed to the Code because it diverted energy away from more pressing issues such as campaigning on arms exports to Indonesia – this was the time of the East Timor crisis, in which UK- and other foreign-supplied weaponry was used by the Indonesian military in a campaign of repression. There was (and still is) a belief within CAAT that if governments wanted to stop exporting, they could, and would not need a code of conduct to do this. For CAAT the role of powerful arms companies in setting the parameters of government policy meant a code of conduct was unlikely to have much effect, according to a campaigner involved at the time. CAAT found itself having to choose between coalition-building with other NGOs and standing alone according to principle and, as explained by former and current campaigners, it chose the latter route. Since then, CAAT has been somewhat marginalised, partly through its own volition, from the wider NGO community. A current staff member described how, once the EU Code was adopted, CAAT altered its position and in November 1998 decided to acknowledge the reality of the Code and work with the coalition to tighten it. But CAAT does little by way of policy work in relation to this, and always points out the bigger-picture arguments as it sees them.

The EU Code came into force in 1998, and its eight criteria include, amongst other things, the commitment not to issue arms export licences if there is a clear risk that the proposed export might be used for internal repression, would provoke or prolong armed conflicts or aggravate existing tensions or conflicts, or be used aggressively against another country (Council of the European Union 1998). Since then the UK Working Group NGOs have been active in trying to promote the strongest possible implementation of the Code, proposing measures and criticising decisions that go against the spirit or letter of the Code (e.g. EU NGOs 2004, 2008). They were pivotal in the push towards the transformation of the Code into an EU Common Position, conducting much of the research that went into the elaboration of the Criteria and advocating for its adoption, which was achieved in December 2008 (Council of the European Union 2008). As a result, the Code is now legally binding on all member states, includes controls on brokering, intangible technology transfers and licensed production, and has expanded and strengthened wording on some points, including, for example, reference to international humanitarian law.

How are we to assess NGO activity in relation to a European arms control regime? NGOs were highly effective in getting their arguments heard by governments, shaping the agenda and moulding the eventual regime. As regional agreements go, the EU Code is one of the tightest and provides a standard against which government practice can be assessed. However, the interpretation of the Code by member states and the ongoing instances of exports that contravene it – documented, for instance, in repeated Saferworld *Audits* of the UK government's annual report on arms export controls – mean it remains 'a form of weak regulatory tokenism – part of a broader process by which all but the most dubious of arms transfers (and sometimes not even those) are provided with a formal veneer of legitimacy' (Cooper 2006b, 121). The effect of the Code's legally binding status remains to be seen, but there is little evidence that a more restrictive approach is around the corner. The Code is thus better understood to serve a legitimatory rather than restrictive function (Stavrianakis 2008). One of the key effects of NGO activity around the Code, and the Code itself, is to reinforce an understanding of the problem of the arms trade as being exports to the South; another is to create the impression of European benevolence. The UK government and arms companies frequently point to the existence of the Code as evidence of restraint, endowing it with significant symbolic power in framing the terms of debate about the arms trade.

Control Arms campaign for an Arms Trade Treaty
There has been a shift in the orientation of the UK Working Group NGO community since the early 2000s, away from policy work on UK export controls and towards international export controls and the community safety agenda in the South, which has a practical focus on small arms. The NGOs have become more internationally focused, using the EU regime as an example of imperfect but best practice as a lever to try to push other states to improve their controls, according to a Saferworld staffer. There is a move away from trying to improve the practices of the EU towards trying to introduce the idea of arms controls in states that are significant in terms of global influence in general and potentially as players in the arms trade, but that are hostile to EU-style agendas, such as Russia and China, and potentially Israel, India and Pakistan. The tactic behind this is to try to start technical conversations about the diversion of military equipment, which most states are keen to prevent, rather than 'tub thumping about human rights', as one Saferworld staffer said. Looking 15 to 20 years into the future, it is the non-European world that is deemed to be the main area of concern with regard to weapons proliferation.

The Control Arms campaign has been a key moment in the inter-nationalisation of NGO perspectives and activity, focusing as it does on getting an international, legally binding Arms Trade Treaty agreed via the UN. The proposed treaty is the biggest international issue in arms control now: there isn't a government in the world that won't have an opinion on it, according to an Amnesty campaigner, although there are limits to NGO influence with some states, such as Egypt, Sudan and Pakistan. According to this campaigner, such states argue that human rights are a stick to beat them with; while their voice is in the minority, it does seem to have a powerful influence on the outcome. NGOs have struggled to counter this argument and have focused, instead, on trying to limit its impact: there is a sense, according to an Amnesty campaigner, that NGOs will not be able to change the views of obstructive states. So they try to emphasise the international nature of the legal principles that the proposed treaty is based on, such as the UN Charter and Geneva Conventions.

The Control Arms NGOs argue that human rights and sustainable development are not 'subjective concepts' but ones to which states are already 'bound by international legal obligations'. 'The vast majority of states recognise this and want an ATT [Arms Trade Treaty] that reflects these obligations' (Saferworld 2009a, 2). Many Southern states, in particular those in Sub-Saharan Africa, use the same language as the NGOs and European states, contributing to the possible emergence of universal initiatives, according to one Amnesty campaigner. The Control Arms position in relation to obstructive states is that a small minority of governments cannot be allowed to block the will of the majority, and that ultimately this is in their interests as well, according to one Oxfam campaigner. In this view, instability and conflict have widespread effects in an increasingly interdependent world. Thus, 'All governments have an overriding interest in establishing a stable and secure international system that saves lives, protects livelihoods and responds to the basic rights and needs of the world's people' (Saferworld 2009a, 4). The aim of the treaty is to create a regime that will generate norms and, over time, pull all states up to a common standard, as an Oxfam campaigner put it.

The UK state is a significant ally of NGOs in the pursuit of a treaty. There was a conscious decision to work with the UK government on the issue of an Arms Trade Treaty, signalling a deliberate attempt at an insider rather than an outsider strategy. As one Amnesty staffer described it, the Control Arms NGOs start from where governments are at, talk to them on their own terms and try to generate change through the presentation of coherent policy proposals and demonstration of mass public support for their demands. The UK government has pledged support for an international arms trade treaty, although not necessarily in the form hoped for by the NGOs. NGOs appreciate that the UK government is mostly focused on getting Russia and China on board so as to create a level playing field for exports, according to one IANSA secretariat member. But the Control Arms NGOs' relationship with the UK government has become more cooperative over recent years, with the government more willing to consult and make changes asked for by NGOs, according to one Saferworld staffer.

As well as UK government support for an Arms Trade Treaty, the Defence Manufacturers' Association, a key industry lobby group, has also come out in favour, welcoming the treaty as a means to 'crack down on illegal traders and proliferators, and also as an instrument to create a more equitable regulatory environment for defence exports worldwide' (Williams 2007). The association believes that any treaty 'would not bring new obligations for UK Industry' (Defence Manufacturers' Association 2006, 4). Industry support for the treaty has been a big boost for the campaign, not least because the industry is very influential with government, according to one Oxfam campaigner. Industry is thus now seen as a stakeholder and an important ally. An Amnesty campaigner described this in similar terms, in that NGOs have been working more strategically with industry as a stakeholder in recent years and have attained a greater appreciation for their perspective on the basis of detailed and intensive discussions. The NGOs acknowledge that there are real industrial concerns which they need to understand. NGOs and industry now have a common agenda, as one Amnesty campaigner put it, related to the harm caused by irresponsible transfers, the negative effect on people of poor export decisions and the role that those decisions have in human rights violations, conflict and underdevelopment.

In welcoming the support of the UK state and industry, the Control Arms NGOs see them as part of the solution to the problem of unregulated arms transfers. While this is strategically effective in terms of getting powerful actors on board and thus ensuring greater likelihood of a treaty, it also perpetuates a number of silences in the debate about the arms trade. It sidelines the question of why these actors sometimes contravene the spirit, if not the letter of their existing obligations; it narrows the definition of a 'controversial' transfer to one that is not authorised by state authorities and/ or is illegal under international law; and it legitimates the scale and scope of the state-sanctioned arms trade.

Call the Shots: CAAT

CAAT has formally signed up to the Control Arms Campaign but does not actively support it. It felt the campaign and any eventual treaty would further raise awareness of the arms trade, act as a step towards greater controls, helping 'prevent sales to human rights abusers, conflict zones and impoverished countries', and be another instrument to help persuade the UK government to adhere to its own criteria on arms exports. However, CAAT's view is that industry's belief that the treaty will not bring new obligations, and the treaty's failure to challenge the impetus to exports, means it would have limited impact (CAAT no date, d; CAAT 2006b, Ev. 162).

This stance is emblematic of the division between CAAT and the UK Working Group/Control Arms NGOs that emerged over the EU Code negotiations. While CAAT has always been opposed to the arms trade *per se*, making it an uneasy partner to the other NGOs, the practical content of its campaigns, up until around 2005, had been broadly similar. For example, earlier campaigns focused on controversial exports to the South, in particular their role in conflict, and the economic subsidy on arms exports. Since 2005 however, CAAT has focused on the relationship between arms companies and the UK government as the impetus to arms exports. Successive stages have focused on exposing the 'revolving door' between companies and government, calling for the closure of DESO, a legal strategy of judicial review against the government's decision to halt the Serious Fraud Office inquiry into allegations of bribery in arms deals with Saudi Arabia, and, most recently, a campaign to 'end the uncivil service' that supports arms exports via UK Trade and Investment (UKTI).

CAAT's strategy in its Call the Shots campaign was to challenge the UK government on key issues such as the revolving door between industry and government, the existence of DESO, and the role of military-related advisory bodies. It did this through public pressure, encouraging supporters to lobby their MPs, and holding an 'action day' in which supporters made a human chain around the DESO building in central London, designating it a 'global danger zone' because of its promotion of arms exports (D'Cunha *et al.* 2007, 8-9). This strategy is unique amongst NGO activity because of its explicit focus on the relationship between state and corporate power. CAAT does not specify whether it wants a re-nationalisation of arms companies, or a re-articulation of the relationship between ostensibly private companies and the state. As a staffer described, making broad vision statements can alienate segments of the campaign's support and so is largely avoided; the focus is on the narrower goal of ending government support for the arms industry. There is ongoing debate within CAAT about the extent to which it should focus on militarisation at home, and how that relates to arms exports and the wider arms trade. To date, however, CAAT has not worked on domestic procurement and militarisation to any degree.

BASIC: Getting to Zero

BASIC was part of the NGO community that contributed to the emergence of small arms on the international security agenda in the late 1990s. It was part of the Biting the Bullet project and the UK Working Group, conducting research and advocacy to tighten UK and international conventional export controls. Since 2007, however, it has returned its focus to transatlantic security, in particular nuclear disarmament. It aims to use its network to strengthen the call being made by 'a group of senior former officials with strong influence within the United States' for 'the complete elimination of nuclear weapons' (BASIC 2007a). While the dominant framing of nuclear weapons is currently through an anti-terrorism prism, BASIC aims to reshape the debate, pursuing technical measures while trying to shift the parameters of debate at an elite level using an insider strategy. Progress on nuclear disarmament is not going to be made until there are some deeper fundamental changes in Western attitudes towards security, according to a senior staff member, and one way of encouraging this is to point to trends and threats that highlight the deep inconsistencies contained within US and UK current strategy. As we see in Chapter Four, BASIC is the most active NGO on UK, transatlantic and European defence policy.

Cumulative impact versus undermining of more radical voices

These brief snapshots of NGO strategies outline some of the key contours of the debate about NGO activity. There are two sets of objectives in play in the NGO world: tighter regulation of the arms trade and the abolition of it. Within this, the various NGOs have different empirical foci: from small arms in the case of the Control Arms NGOs, to major conventional weaponry in the case of CAAT, to nuclear weapons in the instance of BASIC. Again, their strategies both reflect and help shape their arguments. At opposite ends of the spectrum, CAAT's more confrontational strategy both reflects and shapes its more transformist argument, while Saferworld's insider strategy is related to its reformist argument.

The view from one of the Control Arms NGOs is that they have won a lot of the political and ideological arguments. As an Amnesty campaigner said, it is more a technical issue of implementation now, and NGOs are very well placed to influence those discussions directly. One reason for this is that there has been a broad coalition of groups working on arms trade issues for the last decade or so, including big-name agencies such as Amnesty and Oxfam, whose reputation holds sway with decision makers. Coalition work has been a key feature of NGO activism on the arms trade over the years, initially through the UK Working Group and more recently through the Control Arms campaign. This is a deliberate attempt at cumulative impact: as a member of the IANSA secretariat pointed out, Oxfam's research into the opportunity costs of arms transfers combines well with IANSA's focus on guns and Amnesty's research into torture equipment.

Cumulative impact through coalition work has certainly been a key feature of NGO successes on arms trade issues. What is less deliberate and less obvious, however, is the marginalisation of more transformist arguments and solutions through the monopolisation of the space available to NGOs by more reformist organisations. This is evident in two ways: through the marginalisation of particular actors, in this case CAAT, and the competition for issue prominence, in this instance the dominance of human security over inter-state and state–capital relations.

For all their misgivings about outsider approaches, the more insider organisations cite the need for more radical organisations such as CAAT to exist. Campaigners from both Amnesty and Oxfam described CAAT as 'naïve'; working publicly with CAAT is deemed unwise by more insider organisations because of the way it is perceived by policymakers. One Amnesty campaigner understood CAAT to be 'valuable' if 'slightly over the top'. While some constituencies within Amnesty could potentially have much in common with CAAT, overall the organisation relies on an apolitical approach that is at odds with CAAT's ethos. Amnesty's work rests on a 'reputation for informed judgment, rather than random protest' and the role of its researchers is 'not to be too indignant or self-righteous' (Hopgood 2006, 74, 78). In the eyes of some within Amnesty, CAAT falls into both these traps, although there are instances in which individuals are sympathetic to CAAT's arguments even if the organisation cannot publicly endorse them. However, one Oxfam staffer wondered, if CAAT 'didn't demonstrate outside exhibitions, go to shareholder meetings, could we do what we do? Is it because of CAAT that we get access, because we're seen as the sensible end?' There 'must always be a place for ranters' such as CAAT. An Amnesty campaigner felt that CAAT is important because 'government needs to feel there's a public movement completely opposed to the arms trade...CAAT can say 'Stop DSEi' [Defence Systems and Equipment International, a biennial defence exhibition/arms fair held in London] in a way we can't.' Another Amnesty staffer described the UK Working Group NGOs as providing the core of government with workable policy solutions, so the government wants to listen to them. CAAT 'does something different. It's on the outside shouting, while others are on the inside.' In this view, CAAT's role is to create public pressure and awareness, outing bad public decisions. For BASIC, too, complementarity does not need to mean working together; indeed, it can be a disadvantage for insider groups if outsiders move towards the centre of the spectrum. As a BASIC staffer put it, CAAT 'create the political conditions for our insider work to have an effect', and this view is shared within CAAT. There are thus benefits to having outsider groups around in terms of generating policy change.

The protest element of CAAT's strategy is what marks it out from the other organisations. It has a strict code of non-violence that supporters must abide by when participating in protests and direct action organised by CAAT: not only must action be non-violent, it must also 'be seen by those external to CAAT as being non-violent'. This rules out physical violence and carrying 'anything that could be construed as a weapon', but also verbal abuse and 'in some situations...heightened or confrontational language' (CAAT no date, e). While there are a variety of definitions of violence in play in the wider peace movement, CAAT itself tends towards the most restrictive definition. This has brought tension with direct action groups that do not share the same definition, meaning that CAAT treads a difficult line between affirming its code of non-violence, challenging common-sense definitions of violence, and supporting any action aimed at disrupting the arms trade. Internal debates around CAAT's position on the meaning of violence suggest that the organisation directly experiences the issues discussed in Chapter One around the possibility of counter-hegemonic action on the arms trade. Despite CAAT's code of non-violence, the protest element of its work contributes to its being held at a distance by other NGOs.

Overall, though, the view that CAAT creates the political conditions for insider work to have an effect is held both within and outside CAAT. This suggests that the more insider organisations benefit from the presence of more outsider organisations, which raise the political temperature. What is rarely taken into consideration is that outsider organisations are dis-advantaged by the activity of insider organisations. Outsider groups may be rendered ineffective above and beyond their own weaknesses. By seeming reasonable and constructive, the more mainstream agencies monopolise political space available for NGO activity. CAAT looks unreasonable and destructive in comparison, and is thus not taken seriously as a political actor; insider organisations thus reduce the likelihood that outsider organisations can be effective.

This is not, however, just a question of strategy; it is also an issue of complementarity or competition between NGO agendas. That is, CAAT's work can create the political conditions for the insider, yet more transformist work of BASIC, which is focused on geopolitical relations between major powers, while the other insider NGOs focus on humanitarian issues and human security. For BASIC, according to a senior staff member, geopolitics and human security constitute a spectrum, and the latter can be addressed without fundamentally changing the current posture and role of the UK in the world, while the former cannot. Small arms can be addressed as more of a technical problem than can nuclear disarmament, which will require a much greater transformation of political approach. The British government is far more willing to address the human security and small arms agendas because they do not harm its project of retaining status and dominance in the international system; if anything, they enhance it. According to this BASIC staffer therefore, the other NGOs can afford to work much more closely with the UK government. This is borne out in the way in which, while the arguments of BASIC and Saferworld about international security beyond arms exports may well be complementary, their focus and orientation are different: Saferworld focuses on the need to integrate conflict prevention and peacebuilding concerns into development policy, and focuses institutionally on working with DfID; BASIC's wider argument beyond arms exports relates to the preponderance of militarised western definitions of security and military policy.

Overall, then, there is a trade-off between reformist and transformist approaches. NGO suggestions for change have to resonate with existing socially entrenched narratives in order to have purchase. However, while it can be productive to remain on the discursive terrain of less politicised topics, this also further reinforces the existing parameters of debate about the arms trade, keeping more transformist ideas off the agenda. One of the aims of this book is to explore the ways in which NGOs produce and reproduce or challenge the arms trade as an issue in international relations, as well as with the practical or technical effectiveness of their arguments and strategies.

Arms capital integration into the state

As discussed earlier, the liberal fallacy of equating NGOs with civil society can be avoided by contextualising NGO activity in light of the capitalist structuring of civil society. This requires investigation into the role of arms capital and its relationship to the state. Arms capital has been integrated into the structures of the UK state in two main ways: through a 'revolving door' between the state and military industry; and through the high levels of arms company representation on military advisory bodies. This means that arms capital has a structural presence within those parts of the UK state most closely involved in defence industrial policy and arms export promotion. This is in contrast with NGOs, some of which have been integrated (although not to the same degree) with institutionally weaker parts of the state, such as DfID.

The 'revolving door' refers to the traffic of personnel between military industry and the state (in particular the MoD, both the civilian bureaucracy and the military itself), and vice versa. For example, since the start of the 1990s at least three Defence Secretaries and three Defence Procurement Ministers, as well as a number of other MoD staff and senior military personnel, have gone on to be employed by arms companies after their time in public office (CAAT 2005a). Senior figures from the civil service, political parties and the military have gone on to work for arms-producing companies and associated firms such as PR companies. Examples include General Sir Mike Jackson (former Chief of the General Staff, now chair of PA Consulting's Defence Advisory Board), Sir Kevin Tebbit (former MoD Permanent Secretary, now non-executive director of Smiths Group and Chairman of Finmeccanica), Lord Levene (former Chief of Defence Procurement, went on to become Chairman of General Dynamics and President of the Defence Manufacturers' Association). In addition, there is a pattern of secondments between military industry and the MoD (with the traffic going both ways), in part facilitated by the MoD's Interchange Programme. Perhaps the most significant indicator of the revolving door has been that the head of DESO was traditionally drawn from the arms industry and continued to draw pay from companies during his stint there. As is discussed in more detail in Chapter Four, however, DESO was closed in 2007 and arms promotion activity was transferred to the UK Trade and Investment Defence and Security Organisation (UKTI DSO). The current head of arms promotion, Richard Paniguian, came from the oil company BP rather than from an arms company, and the postholder's salary is no longer topped up by industry, as it was in the past. However, arms promotion is still significantly and structurally privileged over other industries in terms of state support.

In addition to the revolving door, arms companies have a significant presence on military advisory bodies such as the National Defence Industries Council (NDIC), through which industry works in partnership with government to set policy priorities. The NDIC is 'the forum in which industry meets the MoD, BERR, UKTI and HM Treasury, alongside Trade Union representatives', meeting four times a year and normally chaired by the Defence Secretary or the Minister of State for Defence Equipment and Support (SBAC no date). Participation in such advisory bodies means that the arms industry has access to very high-ranking officials and politicians in addition to the more general activity of trade associations and organisations that act as liaison between industry and the military. Such organisations include the Defence Industries Council and ADS, the trade body for the UK aerospace, defence and security industries, formed through the merger of the Defence Manufacturers' Association, the Society of British Aerospace Companies and the Association of Police and Public Security Suppliers (APPSS) in 2009. Such bodies help their members build a positive relationship with government, further increasing industry access to elite decision-making processes.

The integration of arms capital into state structures and advisory bodies helps create a pro-arms-industry culture within the state. One of the most important ideological tasks of big business is 'to persuade the population at large that "the business of society is business" and to create a climate of opinion in which trade unions and radical oppositions...are considered to be sectional interests while business groups are not' (Sklair 1997, 526). The most recent Defence Industrial Strategy (MoD 2005), which announced 'sustained real increases in the Defence budget arising from each Spending Review since the government was elected in 1997', is the epitome of the close – indeed overlapping – relationship between the state and military industry. The Strategy seeks to 'share objectives, risks and rewards' between the two so as to be able to 'maintain appropriate sovereignty and thereby protect our national sovereignty' (MoD 2005, 132). The Strategy uses the language of partnership and asks industry to make a parallel commitment for 'planning more effectively and jointly for the long term...including a greater commitment to joint education, staff development and interchange opportunities' (MoD 2005, 11). The language of partnership naturalises the relationship between state and industry, making it seem commonsense and making the demands of NGOs appear as special interests.

This is not to argue that the interests of the state and capital are identical, or that only policies beneficial to the military and the arms industry come into operation. There are frequent and vigorous disputes between fractions of the state and capital, often over domestic procurement issues, and large companies claim to be discriminated against by excessively stringent UK export guidelines and perceived lack of government investment and support (see, for example, Hope 2004; Jameson 2006; Morris and Khan 2005). Rather, there is a structural bent towards pro-military and pro-industry policies. More generally, arms exports are legitimated by historically entrenched claims of national defence. The argument that arms exports are politically, economically and strategically beneficial carries weight above and beyond its factual content because of the underlying ideological function of the term 'defence', and claims regarding state sovereignty serve as a powerful justification for the very idea of an international trade in weaponry. Military activity and the arms trade have an almost sacred quality even among those elements of the state that are not embedded with arms capital (Melman 1970, 26; Chomsky 2003, 71; Kurtz 1988, 53, 95).

Labour, as represented through trade unions, is weakly or minimally integrated into decision-making structures and therefore has limited influence compared to the integral role of industry in policy planning. Whilst trade unions are represented in bodies such as the Aerospace Innovation and Growth Team (AeIGT) and the NDIC, their number, seniority and weight is limited compared to that of arms capitalists. More significantly, however, trade unions call for fundamentally similar measures as arms capital: given the mandate of trade unions to protect the jobs of their members, it is unsurprising that the unions call on the government to dedicate more resources to the arms industry. For example, Amicus (the largest union to represent arms workers) is in favour of Export Credits Guarantee Department (ECGD) insurance cover on arms deals, because the support facilitates orders and therefore protects British jobs; it also recommends an increase in defence spending (Wall and Johnston 2004). It argues that British workers are discriminated against because European companies are often at least part-owned by the state and the US state subsidises research to a greater degree (Amicus 2004; Unite no date). Rather than trade unions arguing against arms capitalists and the state, therefore, we see them arguing for essentially the same thing. There is a significant degree of cross-class support for military production (Melman 1970, 225), with the effect that it becomes difficult for critiques to be articulated that are simultaneously anti-export (or particular exports) and pro-worker.

Rethinking global civil society: dual networks

Paying attention to the structural positioning of NGOs, arms capital and labour means we need to rethink our conception of civil society and hence our assessments of the potential for NGO effectiveness. Rather than conceiving of the state, market and civil society as three spheres, as is commonly done in the mainstream liberal literature, the conception put forward here is of dual networks that comprise actors from across government, capital and NGOs. This helps us understand the ambiguity of civil society and better appreciate both the constraints and opportunities for NGOs.

Arms capital has been integrated into the structures of the UK state with the effect that the parameters of defence industrial and arms export policy are set by an elite group of state (in particular the MoD and BIS) and industry actors, with the latter being dominated by BAE Systems in particular. This significantly weakens the relative autonomy of the state and forms the core of a military-industrial complex (Eisenhower 1961; Lens 1970; Rosen 1973; Sarkesian 1972) that operates in favour of higher military spending, support for arms companies and arms exports. Meanwhile, NGOs have made alliances with DfID, in particular its Conflict, Humanitarian and Security Department (CHASE, formerly CHAD, Conflict and Humanitarian Affairs Department), but also those elements of the FCO and MoD concerned with preventing small arms proliferation. Amongst the NGOs, Saferworld enjoys the closest relationship with the state.

Those elements of the state with which NGOs are allied are institutionally weaker than those with which arms capital is integrated, in that they lose out when hard interests are perceived to be at stake. For example, in the case of the controversy over the granting of an export licence for an air traffic control system to BAE Systems for export to Tanzania in 2001, the licence was granted despite the opposition of DfID, the Treasury, the World Bank and the IMF, all of which are involved in Tanzania's aid programme. Key ministerial interventions (including, it seems, from Prime Minister Tony Blair) led to the granting of the licence despite the arguments against it. It has since been revealed that allegations of bribery have been made against BAE Systems in this case which, if true, would raise questions about the knowledge the UK government had of this and the reasons for its continued support for the deal. In this case, and in the debates about developments and arms exports subsequent to it, NGOs called for DfID's concerns in arms export licensing to be taken more seriously. However, regardless of their strategy, NGOs were marginalised as political actors in this case and DfID was powerless to prevent the deal (see Stavrianakis 2005).

NGO influence can thus in one sense be understood as 'less' than the influence of arms capital, in that NGOs do not have as much institutional weight or leverage as arms capital in setting state agendas and are ignored when key strategic and economic interests are perceived to be at stake. However, DfID exercises considerable power in terms of the conflict, security and development nexus, and NGOs have been highly influential in developing this, as is argued in more detail in Chapter Six. Given that the conflict, security and development agenda is now a hegemonic form of international action in terms of North–South relations (Duffield 2001, 2007) and is central to UK foreign and development policy, NGOs are considerably more powerful than we might at first think.

Thus, understanding civil society as riven by the inequalities of capitalism requires us to do two things: to appreciate the disjuncture in structural position of NGOs, and also to think about the ways in which they construct the arms trade as a problem in international relations, and ask whether this is transgressive or not. While no articulation is inevitable or necessary (meaning any articulation 'can potentially be transformed'), what is at stake is the question of whether NGOs can rearticulate the problem of the arms trade 'to break, contest or interrupt some of these tendential historical connections' (Grossberg 1986, 54). Overall, the activities of these dual networks are complementary rather than antagonistic, just as military production and trade, and arms control are complementary. They demonstrate the ongoing significance of coercion in backing up state and capital, and the simultaneous operation of both imperial and liberal forms of power.

CAAT is an exception to NGO integration into state-led networks, due to its more confrontational strategy and its arguments, which fall outside the accepted parameters of mainstream political discourse. This does not mean that CAAT is left to its own devices, however: it has been subject to considerable political interference, including surveillance, infiltration and spying. In the mid to late 1990s, when CAAT was campaigning against the export of Hawk jets to Indonesia, it was infiltrated by at least half a dozen agents of a private intelligence company, Threat Response International, run by Evelyne Le Chene. The agents collected personal and campaign information, some of which was then sold on to BAE Systems and used to pre-empt protest actions, spread disinformation and manipulate campaign activity (CAAT 2005b; Insight 2003; Lubbers and van der Schans 2004). Thus, when CAAT instructed solicitors to seek judicial review against the government for its arms exports to Indonesia, 'BAE was alerted to the contents of a letter sent by the firm to the then trade minister, Ian Lang', and a letter by Foreign Office minister Jeremy Hanley to CAAT regarding arms sales to Indonesia was also obtained by BAE, which then used the information to remain one step ahead of campaigners when lobbying in parliament (Insight 2003).

This was not the end of the matter, however. In 2007, as CAAT was launching a case of judicial review against the UK government in light of its decision to terminate the Serious Fraud Office (SFO) inquiry into allegations of bribery against BAE Systems in relation to arms sales to Saudi Arabia, privileged legal advice from CAAT's solicitors was obtained by BAE Systems (CAAT 2007a; Monbiot 2007). CAAT's legal action in relation to this discovery led to the revelation that BAE Systems had been paying £2,500 per month to a company called LigneDeux Associates to monitor CAAT and pass information to the company. It also emerged that BAE Systems had been involved in spying on CAAT – unlawfully – to a greater degree than originally implied; the company later admitted for the first time that it had employed Le Chene to monitor CAAT and pass on information (CAAT 2007a).

Thus, while CAAT is not integrated in the same way as other NGOs, it is still enmeshed in state–capital networks, albeit involuntarily. The effect of this is twofold: there is an obvious negative effect on CAAT's campaigning, as arms companies (and possibly the state as well, given the relationship between them) know what they are planning; and infiltration creates suspicion and mistrust within campaign groups, a likely secondary aim of such activities. However, that a small, under-resourced, non-violent campaign group working out of a shabby office in north London warrants extensive surveillance by one of the world's largest arms companies is itself an indicator of effectiveness, suggesting that the organisation poses a threat above and beyond its physical resources. In addition, the revelations have been an embarrassment for the government and BAE Systems, further denting their reputation.

Conclusion

NGOs are agreed that the arms trade poses a problem in world politics, but they differ in their understandings of the nature of the problem and the most appropriate responses to it. The NGO community is home to reformist and transformist visions, as well as insider and outsider strategies, which sit in uneasy and not necessarily complementary relation to each other. The picture is further complicated, however, when we situate NGO activity in a broader context of the dual networks active on arms trade.

The reformist vision and mix of insider and outsider strategies of Amnesty, Oxfam, Saferworld, IANSA and International Alert are complementary, often through formal alliance. The transformist vision and insider/outsider strategies of BASIC and CAAT are not explicitly allied, but are complementary. While NGO workers are conscious of the need for complementarity of approaches, with outsiders keeping up political pressure for insiders to make headway, what they rarely consider is that reformist visions can undermine transformist ones. So insider and outsider strategies *can* be complementary, but outsiders are marginalised unless the insiders are also transformist. And all of this must be considered in light of the structural disciplining of the NGO sector, in terms of its relative weakness compared to state–capital relations and the liberal basis of much of its analysis. The following three chapters explore in more detail the argument set out thus far, exploring NGO activity in relation to the three main trends of the arms trade: intra-Northern production and trade; North–South transfers; and small arms proliferation.

4 • Arming the North: Transatlantic and European Military Production and Trade

Military production and trade within the North are key components of contemporary globalisation. The post-Cold War, post-9/11 international security landscape is widely understood to feature a new range of threats, such as terrorism, new wars, failed states, and the illicit drug and small arms trades, which require active intervention. In this view, new forms of military doctrine, strategy and tactics are needed, along with new or at least adapted forms of military equipment and procurement (see, for example, Arquilla and Ronfeldt 1997; Gates 2009). And accelerating processes of economic globalisation are deemed to make the high-skilled jobs and economic competitiveness associated with military production increasingly important to states (e.g. Hensel 2008; MoD 2005).

NGO activity in relation to military production and trade within the North is focused on a few specific issues. BASIC, CAAT and Saferworld have intervened in debates around the economic arguments in relation to UK arms exports, focusing on the economic subsidy, the jobs argument and the role of particular institutions that support arms exports, such as the Defence Export Services Organisation (DESO) and the Export Credits Guarantee Department (ECGD). The NGOs have done a smaller amount of work on intra-Northern defence collaboration, commenting on the impact of subsidies on domestic procurement, the risks to the UK of asymmetrical defence-industrial collaboration with the US and the export control risks associated with the Europeanisation of defence production.

NGO activity on intra-Northern production and trade has diverged over time. In the late 1990s and early 2000s, there was concentrated NGO effort at tackling the economic arguments around arms exports, which was transgressive and may have borne fruit through the closure of DESO in 2007. In this sense there has been a significant degree of complementarity of NGO activism. However, since around 2003 there has been a split in priorities within the NGO community. Saferworld has shifted its focus away from national UK policy towards EU-level and international action aimed at an international Arms Trade Treaty, as part of the Control Arms campaign. CAAT and BASIC, meanwhile, have continued their focus on national and transatlantic policy and practice, respectively. Despite this difference, it is notable that the main way in which transatlantic and European defence integration is articulated as a problem is as one of export control standards. That is, intra-Northern military production and trade, and defence integration within the European world are not understood by NGOs as security problems in and of themselves. Rather, they are articulated as defence-industrial issues, the negative security ramifications of which arise from exports to the South.

Minding the purse strings: the economics of arms exports

Economic arguments are central to mainstream positions in favour of arms exports. The UK government calls the arms industry 'outstandingly successful and a vital national asset' (MoD 1998, ch. 8) and describes it as 'an important part of the UK economy, contributing billions of pounds in exports each year, and providing tens of thousands of jobs in the United Kingdom' (DSO 2009). The main economic claims made in favour of military exports are that they generate jobs, reduce MoD procurement costs and contribute to the balance of trade. This position is a key weapon in the ideological debate over arms exports and finds cross-class support across the Labour and Conservative Parties, arms companies, other business lobby groups and trade unions (e.g. BAE Systems no date; Clarion Events no date; Defence Industries Council 2009; DSO 2009; Howard 2005; Unite, no date). Criticisms of UK arms export policy are frequently sidestepped with an appeal to this supposed national asset. For example, when faced by a journalist's question on the ethics of arms exports to India during the Kashmir crisis of 2002, Prime Minister Blair responded that 'The idea that we shut down our defence industry in these circumstances I find bizarre' (Blair 2002). Even though every one of the government's claims about the economic (and other) benefits of arms exports is open to challenge (see Mayhew 2005), the ideological terrain is weighted against NGOs seeking to challenge the supposed economic benefits of arms exports.

BASIC, Saferworld and CAAT have been the most active in challenging the economic arguments around arms exports. Common to their different strategies has been the tactic of conducting in-depth research to provide a factual challenge to the UK government. For example, they calculated the economic subsidy on arms exports, with BASIC and Saferworld arguing that arms exports are subsidised by at least £420m per year, through export credit guarantees, marketing and other government support, tax breaks on bribes and other corrupt practices, and distortion of MoD purchasing and other priorities. In addition, government research and development generates an extra subsidy of up to £570m per year (Ingram and Davis 2001). CAAT, calculating the subsidy somewhat differently, claims the figure is as high as £888m, of which £670m is research-and-development-related subsidies. While their figures differ, CAAT also points to export credit guarantees, distortion of MoD purchasing, DESO support, and the use of embassies and defence attachés as key elements of this support (CAAT 2004).

As a result, all three NGOs claim that the employment benefits of the arms industry are overstated. According to BASIC and Saferworld, employment from arms exports accounts for only 0.3 per cent of UK employment, and each job directly related to arms exports is subsidised by £4,600 per year (Ingram and Davis 2001; see also Ingram and Isbister 2004). CAAT's latest research demonstrates that the 55,000 jobs related to arms exports and 155,000 jobs related to producing equipment for the UK armed forces, taken together, constitute less than 0.7 per cent of the UK workforce and around 7 per cent of manufacturing jobs (CAAT 2009a). Jobs dependent on exports constitute an even smaller percentage – less than 0.2 per cent of the UK workforce and less than 2 per cent of manufacturing jobs, and these jobs are subsidised by at least £9,000 per job per year (CAAT 2009a).

The factual basis of the NGOs' research is similar, but they deploy it towards different strategies. BASIC and Saferworld engaged in a strategy of trying to shift the government from where it is to where they want it to be through a process of negotiation. For example, they emphasise that 'we are not calling for the defence industry to be closed down, an accusation that is frequently made whenever commentators question government support for particular arms exports' (Ingram and Isbister 2004, 9). Rather, they are concerned that the government's claims 'have not been backed up with clear economic evidence' (Ingram and Davis 2001). Thus, while exports are undoubtedly good for arms-producing companies, they come at 'an undetermined cost to the economy at large', and the NGOs seek to 'challenge the misuse of public resources' (Ingram and Davis 2001, 6, 2).

These public reports were backed up with insider advocacy by BASIC, mostly based on personal connections with those within government. Through building relationships with the Treasury, the Prime Minister's

Office and the ECGD in relation to the academic debate, and through a strategy of highlighting the inconsistencies and weaknesses of the dominant arguments usually given in support of arms exports, BASIC's staff felt it was able to generate 'something of a bushfire within the Treasury' in relation to subsidies. In this view, economic justifications for arms exports are 'intellectually bankrupt' but they have 'political clout', which means the presentation of information is as important as the content. BASIC operationalised a similar strategy in relation to DESO, working closely with the Treasury and engaging in behind-the-scenes work with 'officials at the heart of government' involved with the decision to conduct a comprehensive review of DESO's functions (BASIC 2007b).

CAAT, in contrast, does not feel the need to issue reassurance that it does not want to close down the industry, and does not start from where the government is at. It argues that the government subsidy 'completely undermines the "jobs" argument that is routinely used by government to justify arms deals. If politicians wish to promote military exports, they should be open about their motives rather than hide behind an economic fiction' (CAAT 2002a). For CAAT, 'the lack of a significant economic benefit for the UK removes the standard justification for the arms trade. As the subsidy is actually several hundreds of millions of pounds, it provides a positive economic reason to stop the trade and, given the political will, the resources to help that happen' (CAAT 2004). In recent research, it makes the case for possible 'alternative and useful jobs' – in renewable energy, for example – that could be created by the government if it had the political will to move away from arms production (CAAT 2009a). CAAT focuses on the jobs argument because it serves as the government's 'fallback position', according to one staffer. That is, 'The Treasury don't buy it but the public believe it and the government and obviously the arms industry want them to believe it, so we need to challenge it.'

To challenge such arguments, one element of CAAT's work is to focus on educating the public, so as to create pressure on the government. In 2002, it ran a campaign called 'Shelling Out', focused on the way taxpayers' money is used to subsidise the arms trade. This was intended to give supporters information to counter the government's arguments. Although it did not generate as much support as other, more emotive campaigns that focused on the role of UK-supplied weapons in conflict, or the impact of the arms trade on children, the work on economic arguments generated the momentum for CAAT's future focus on the relationship between arms companies and the government. The view within CAAT was that state–capital relations were the driving force behind the UK arms trade. Within the wider NGO community, there was anecdotal discussion of the links between government and industry, but no other NGOs were actively researching it or campaigning on it, reinforcing a feeling within CAAT that it was urgent to work on this.

CAAT's public education efforts complement its policy work with MPs and officials. For example, it briefed the then Financial Secretary to the Treasury, Stephen Timms, on the economics of the arms trade in spring 2007. This process included a meeting with Fellowship of Reconciliation and Speak, two religiously oriented organisations that were also campaigning for the closure of DESO, and whose presence with Timms (described by CAAT as having a strong Christian philosophy) was felt to be important. CAAT had also previously, in 2006, engaged in a popular campaign to 'Shut DESO,' which included a petition with over 10,000 signatures, postcards to more than 400 MPs from constituents, an action day that designated the DESO headquarters as a 'global danger zone' with a cordon of protesters around it, as well as news articles in the media and a statement calling for DESO's closure, supported by the Liberal Democrats, Plaid Cymru, the Green Party and more than 30 other NGOs (CAAT no date, f).

These different types of NGO activity paid off in 2007 when Prime Minister Gordon Brown announced the closure of DESO and the transfer of responsibility for defence trade promotion to UKTI (Brown 2007). The announcement generated anger from the arms industry, with BAE Chief Executive Mike Turner writing to Gordon Brown to express his 'considerable concern' about the announcement and the 'complete lack of consultation with the industry stakeholders', which '[seems] quite contrary to the intent set out in the Defence Industrial Strategy document and seems to be out of keeping with your own statements about consultation with the business community' (Turner 2007).

Industry anger over the decision is an indicator of NGO success in and of itself, but the question of how significant a change the closure of DESO is remains a live one. After all, the decision was a 'machinery of government' change rather than a policy change, aimed at integrating trade promotion for defence exports more effectively with the government's wider trade support activities. Existing and planned government-to-government deals will continue to be administered by the MoD, and those DESO functions that support UK defence policy will be moved elsewhere in the MoD (Brown 2007). In April 2008 the Defence and Security Organisation was established within UKTI; half of DESO's nearly 500 staff moved over to

UKTI and there was a small reduction in staff numbers, although DSO still has as many staff as the other UKTI sectors put together (CAAT 2008a). Overall, government support has moved from a dedicated government department to a sub-section of a department, and seems to be slightly less obvious. However, the Conservative Party has pledged to return DESO to the MoD, should it return to power (Barker and Parker 2008; Fox 2009).

How are we to assess NGO effectiveness in this instance? CAAT and BASIC claim success for the DESO closure (BASIC 2007b; CAAT 2008a) and Saferworld had been involved in the longer-term efforts to chip away at the economic rationales for arms exports. The material impact in terms of financial and political government support for arms exports remains to be seen. But the closure of DESO also has great symbolic value. As Saferworld had earlier put it, the value of DESO is 'not just monetary' – it gives, 'in effect, an official stamp of approval to prospective arms sales' and 'relentlessly promotes the industry's interests within the government' (Saferworld 2007b, Ev. 74). The closure of DESO and the transfer of some functions away from the MoD is an indicator of potentially transgressive change. However, as government-to-government deals and some other tasks remain the responsibility of the MoD, the significance of this shift can be questioned.

NGOs are conscious of the need to ensure that the change at least opens up the possibility of challenging government support for the arms industry. BASIC wants to 'ensure that the closure of DESO heralds a deeper change in government policy and not just a re-organisation of the same activities from one agency to another' (BASIC 2007b). Saferworld emphasises the importance of ensuring 'that the defence industry is no longer privileged at the expense of other sectors of the economy' (Saferworld 2007c), and CAAT hopes that the closure of DESO will mean that 'the undemocratic power of arms companies in the UK will be brought to an end' (CAAT 2007b). It followed up on this by launching a campaign to 'end the uncivil service' in which UKTI is deemed 'armed and dangerous' (CAAT 2008b). CAAT calls for staff numbers in UKTI Defence and Security Organisation to be 'reduced so that the military equipment sector's exports are given no greater share of UKTI resources than is proportional to the size of the sector's contribution to the economy. Unless this happens, the arms industry will continue to be an undeserved special case' (CAAT 2008c).

NGO work on the economic arguments around arms exports is a good example of the complementarity of different NGO strategies, although there is no evidence that the NGOs deliberately coordinated their campaigning in this way. That is, the outsider strategy of CAAT helped create the public pressure for insider advocacy, in particular that of BASIC, to have effect. This view is echoed within the NGO world: Saferworld and BASIC staffers point to the Labour Party's traditional fear of being seen as an enemy of business and weak on defence, which makes it imperative, strategically speaking, to deal with issues as technical rather than political matters. These NGOs benefit from outsider groups taking a no-compromise position: having campaign groups such as CAAT calling for more radical measures allows insider organisations to do their work more effectively, according to a BASIC staffer. In this view, DESO would never have shut if it was not for ongoing CAAT pressure, but it was not the CAAT pressure that shut DESO by itself. Complementarity thus does not necessarily mean working together.

The closure of DESO is a good example of the cumulative effect of different NGO strategies. The work of BASIC and Saferworld proceeded on the basis of an insider strategy and challenged arguments from within the dominant paradigm, strengthening the hand of those within government that were unhappy with the status quo. An incremental reform, such as getting Treasury officials to rethink how they calculate the subsidy on arms exports, can bear transgressive fruit. Given the entrenched assumptions in mainstream political debate that the arms industry is good for jobs, working within the technical realm can be an effective way to generate change. In terms of the way forward, BASIC's and Saferworld's strategy is to let economic arguments wither on the vine, while CAAT continues to challenge the jobs argument as part of its wider remit of promoting progressive demilitarisation in arms-producing countries.

This implicit alliance between BASIC and CAAT marks them out from the other NGOs. Saferworld has moved more towards the Control Arms campaign, broadening its focus away from the UK government and towards generating international momentum for an Arms Trade Treaty. And the economic arguments around UK arms exports are not much of an issue for Oxfam and Amnesty. While the issue of the subsidy sometimes generates knock-on effects for the debate about sustainable development and the arms trade, in the view of a BASIC staffer, Oxfam and Amnesty have not got very involved in it because it does not fit well enough with their respective development and human rights angles. While the other organisations are happy for BASIC and CAAT to work on these issues, and are supportive of them, economic arguments around UK involvement in the arms trade have not been a core issue for them.

Intra-Northern military production and trade

Intra-Northern production and trade – arms purchases by the UK government and exports to the UK's main allies (NATO and EU member states) – are generally not an issue for NGOs other than BASIC and CAAT, and even for them it makes up a limited part of their overall work. Within Amnesty, there is what one campaigner described as 'an ongoing internal debate' about arms sales to the UK military and NATO allies. The overall view is that the UK industry is seen primarily to support NATO and the UK army, with sales that tend to be of high-tech equipment that is unlikely to be used in human rights violations. As this campaigner described it, violations by NATO forces that occur in the course of military operations are treated through the usual Amnesty channels for criticising human rights violations, rather than as an arms control issue. That is, the arms transfer team sees everything through the lens of export controls: there has to be an international transfer, otherwise it is an issue of corporate social responsibility. For example, Amnesty would be extremely critical of the use by NATO of a high-tech weapon to bomb a building indiscriminately, killing civilians. But the export control team would not comment on this if it was done by the UK military as it is a human rights issue, not an arms trade issue.

To use another example, Amnesty called on the US-led occupying powers in Iraq to respect human rights and international humanitarian law (e.g. Amnesty International 2003a), and criticised the indiscriminate effects of weapons such as cluster bombs, calling for an immediate moratorium on the use of these and other indiscriminate weapons (Amnesty International 2003b). But the use of cluster bombs is not used to illustrate the need to control the arms trade, because it does not involve international transfers. US forces' use of other, supposedly more discriminatory military equipment, such as Apache helicopters, small arms and missiles in human rights violations in Iraq (examples from Amnesty International 2003b), is not used as an argument for tighter restrictions on exports to the US from other arms-producing countries, or to back up calls for limitations on US military production. One strand of the arms trade – intra-Northern military production – is thus sidelined as a problem.

In its annual audit of the UK government's reports on arms exports, Saferworld uses a methodology that excludes consideration of arms transfers within the North on principle, despite the intention of the audit to 'influence the development of a restrictive interpretation of the criteria' (Saferworld 2002a, 42). Recipient destinations are split into three categories: 'non-sensitive countries (members of the EU, OECD and most NATO countries which are generally regarded as "safe" destinations)'; 'sensitive' ones that 'appear to "trigger a concern" under one or more of the eight criteria'; and 'very sensitive' ones, that is, those states under embargo. Audits focus on the latter two categories, with licences granted for military transfers to states in those categories being scrutinised against the Consolidated Criteria (as the eight criteria of the EU Code are known in the UK). Further, as 'The granting of an arms export licence depends on two overall factors: the nature of the recipient and the level of sensitivity of the arms, goods or technologies being exported' (Saferworld 2005, 91), Saferworld is following the government's cue in focusing on the more sensitive recipients and technologies, a practice that has long been the general operating principle of the UK Working Group on Arms (UKWG) and other NGOs as well. So domestic procurement is off the agenda for the NGO community, and intra-Northern trade is not an issue of concern in and of itself. However, the NGOs do a limited amount of work on the knock-on effects of subsidies for UK military procurement, the potential risks around defence collaboration, in particular asymmetry within US–UK relations, and European defence integration, both of which are articulated in terms of the danger of lowering standards for North–South exports.

UK domestic procurement and military posture

One issue that emerges from NGOs' arguments about the economics of arms exports is the effect of subsidies on UK procurement and military posture. Saferworld argues that the government's habit of linking export orders with domestic procurement 'can lead to less suitable and more expensive equipment for our armed forces' in that it 'sometimes leads the MoD to purchase kit that is inferior or more expensive than that available from non-UK sources' (Saferworld 2004a). For example, in 2003, as part of an effort to persuade the Indian government to purchase Hawk jets, produced by BAE Systems, the Secretary of State for Defence ordered that Hawks be procured for the RAF in order to demonstrate their desirability. This was against the wishes of military leaders and the Treasury, who preferred a better-value, technically superior Italian-produced aircraft (Tomlinson 2003; Gow 2003). Saferworld claims that 'rather than reducing the cost to the UK, it actually added a £1bn price tag to UK taxpayers'. In response,

Saferworld calls for the government to 'separate defence procurement from export decisions in order to ensure that procurement decisions are taken on the basis of defence needs and value for money' (Saferworld 2004a).

BASIC and CAAT point out the cost and time over-runs of domestic procurement projects, emphasising that they all feature BAE Systems. It is the prime contractor on the five projects running most over budget; it is also likely to supply the replacement for Trident nuclear missiles (Ingram 2006). Astute submarines are likely to cost the government 47 per cent more than originally intended, and Type 45 destroyers are likely to cost 18 per cent more, 'costing the taxpayer £2.2 billion more than expected' (Prichard 2008). More generally, they emphasise the 'monopoly position' (Schofield 2008) and 'growing stranglehold...on procurement' exercised by BAE Systems, and the influence of 'a powerful military industrial network' on military posture that emphasises high levels of military R&D and procurement and an aggressive arms export policy (BASIC 2006b; Schofield 2006). As a result, the UK defence industrial strategy 'conflate[s] the interests of a private company with the interests of the country' and means the UK is 'a pocket-superpower so tied to US military strategy that it is incapable of making a rational analysis of its own security needs', continuing to define security in narrowly military terms despite the end of the Cold War (Schofield 2008, 2006). This argument is shared by both BASIC and CAAT, in that several of their reports are authored by the same person, working as a freelance researcher and peace research consultant.

The financial crisis that began in 2007 has demonstrated another opportunity for informal complementarity of NGO work, and a potential moment of transgression in terms of domestic procurement. For BASIC, it provides a window of opportunity for low-key, technical insider advocacy. According to one staffer, the economic destruction capitalism is currently experiencing could mean that the government will not have the money to fulfil its plans for more carriers and frigates. This provides an opportunity to use cost arguments rather than political arguments in favour of dropping big defence projects. This has always been the route taken by BASIC in its work on the issue of Trident replacement. In its engagement with the parliamentary committee, for example, it called on the government not to make a decision based on 'assumptions from outdated debates of the 1980s' and not to heed the exaggerated claims being made by industry. Rather than calling explicitly for the abandonment of Trident, it argues that a decision on replacing it 'can be delayed for a further eight to 10 years' (BASIC 2007c). Again, this demonstrates a strategic choice by BASIC to keep economic arguments about the arms trade in the technical realm so as not to raise political opposition, which is especially important with a Labour government apprehensive of being seen as soft on defence. Following the financial crisis, CAAT's response has been to argue that the government's response to the crisis has demonstrated that 'politically it is entirely possible for the government to intervene in individual sectors of the economy, and with colossal sums of taxpayer money'. As such, the government could also redirect investment away from defence and towards highly skilled jobs in capital-intensive sectors such as renewable energy, or towards more labour-intensive energy efficiency programmes (CAAT 2009a).

In response to the contemporary orientation of UK military posture, BASIC and CAAT outline similar visions of an alternative defence and security policy. This could see the UK 'play a leading role in the development of the EU as an independent power in world politics and international security', or even replacing the Defence Industrial Strategy with 'an International Security Industrial Strategy (ISIS)' to focus on climate change and other threats (BASIC 2006b). Such a use of 'soft power' would include a shift towards territorial defence, international peacekeeping and reconstruction for UN-endorsed humanitarian interventions (Davis 2007). BASIC calls for NATO to return to a focus on collective defence, rather than commit itself further afield geographically and in terms of mission. It argues that NATO should adopt a broader concept of security, accept that not all security threats require a military response, eliminate battlefield nuclear weapons from NATO, adopt a non-nuclear weapon security doctrine, and improve transparency, accountability and value for money within NATO, the current lack of which is 'symbolic of a democratic deficit at the heart of the Alliance' (BASIC 2007d).

Writing for CAAT, Schofield goes further and calls for a Non-Offensive Defence (NOD) policy, a fundamental shift in UK military doctrine that would involve a move away from the long-standing post-World War Two stance of 'support[ing] the United States in global force projection' as a 'significant but dependent military ally' (Schofield 2007, iv, 16). The shift to NOD would focus on 'territorial defence and a contribution to an EU peacekeeping and reconstruction force that can carry out UN-endorsed humanitarian interventions'. It would involve the cancellation of certain major offensive platforms such as Trident and move in favour of 'a new international security architecture based on global disarmament' (Schofield 2007, iv) as well as an increased emphasis on economic and environmental, over military security (Schofield 2008, 22-3).

While the other NGOs would likely agree with some of these scenarios, what is notable is that they do not articulate this wider vision. As we will see in Chapter Six, Saferworld, Oxfam and International Alert try to integrate concerns about small arms proliferation and conflict prevention into mainstream development thinking, but only BASIC engages in any policy work on trying to transform defence policy. Overall, in terms of strategy, on the question of domestic procurement and UK military posture we see a similar pattern as in NGOs' work on economic arguments. Saferworld makes criticisms from within the government's own parameters, while BASIC engages in a dual strategy of insider advocacy combined with a wider public transformist vision. This vision is shared by CAAT, which does not engage in as much policy-level work but rather focuses on public education to back up its work with MPs and officials, in order to try to create pressure for more fundamental change to occur.

Asymmetrical UK–US relations

BASIC and Saferworld have both engaged in efforts to improve collaboration, information exchange and practices between the US and the UK as part of defence collaboration. In this mode, they are supportive of transatlantic defence collaboration, but try to protect the UK's ability to exercise control over its exports. For example, BASIC and Saferworld intervened in the debate over the 2007 UK–US Defence Trade Cooperation Treaty, which is aimed at 'reducing barriers to the exchange of defense goods, services, and information' in pursuit of interoperable forces and collaborative defence capabilities (US Department of State 2007; also MoD 2007). This proposed streamlining measure permits certain exports of defence materials and services from the US to the UK without the need for export licences, and affirms the UK policy of the use of open licences for exports from the UK to the US, which allow unrestricted quantities of certain types of equipment to be exported to a named end-user.

The two NGOs argue that the treaty is asymmetrical, prioritising US interests over UK ones, as the US would have discretion over what items are covered by the treaty, a right of veto over which UK parties are covered by it, and the right to monitor end-use of weapons developed under the treaty, and would not need UK permission to re-export goods, while the UK would need US permission to do the same (Saferworld 2007b; BASIC and Saferworld 2007). They argue that the treaty 'may well have the consequence of eroding UK operational sovereignty' and risks complicating the UK's relationship with other EU states in terms of the implementation of the EU Code of Conduct (BASIC and Saferworld 2007). They also argue that it risks undermining UK arms transfer controls by subordinating them to those of the US, which has a greater focus on foreign policy and on national interest concerns in arms exports, as compared to the UK's emphasis on human rights (Saferworld 2007b).

A further example is the dispute over what is known as the 'ITAR waiver'. The International Traffic in Arms Regulations (ITAR) are US government regulations for the import and export of military and related equipment and services, and contain restrictive measures such as a prohibition on the re-export of US-supplied military equipment without express US government authorisation. Canada has historically been exempted from the ITAR, and Australia and the UK have attempted to negotiate exemptions from it in order to streamline defence industrial cooperation and military interoperability. However, Congressional concerns over the possible re-export of defence technologies to US adversaries have blocked ratification of the waiver. Saferworld and BASIC argue that preferential licensing arrangements such as an ITAR waiver should be used to raise the export control bar, otherwise they risk 'increasing proliferation and the incidence of irresponsible arms transfers', an outcome merited by neither the demands of the War on Terror nor defence industrial interests (Davis and Isbister 2003, 6). They recommend that the US insist that to qualify for preferential licensing, participating states 'bring their own export controls into line with "best-practice" in the United States' (Davis and Isbister 2003, 7). In the UK context, they argued against proposals to privatise the Export Control Organisation, which manages the export licensing system, arguing that this would further confirm US suspicions that the UK system is unreliable and make Congress even less likely to grant an ITAR waiver (UKWG 2006).

The dispute over the ITAR waiver is an issue of asymmetry because the US refuses to be bound by other states' re-export clauses, while insisting that others respect its demands. This has been highlighted in what has become known as 'the incorporation issue'. In July 2002 the UK government announced a change to its arms export guidelines and its methodology for assessing licence applications for components that are to be incorporated into military equipment for onward export. In addition to the criteria set out in the EU Code, the government now also assesses licence applications against factors such as 'the importance of the UK's defence and security relationship with the incorporating country' (Straw 2002). This policy change was made in direct relation to the export of components for head-up display units to the US for incorporation into F-16 fighter jets and onward export to Israel. At the time Foreign Minister, Jack Straw, stated that, while the UK-supplied content of F-16s is less than 1 per cent, the supply of head-up display units 'is part of a long-standing collaboration in this US programme. Any interruption to the supply of these components would have serious implications for the UK's defence relations with the United States' (Straw 2002).

In response, Saferworld argued that the new guidelines 'appear to open a significant export licensing loophole and a contradiction at the heart of UK policy', in that they have facilitated the export of equipment that would be unlikely to have been awarded a licence for direct export under the Consolidated Criteria (Saferworld 2003a, 9). It claims that the new guidelines provide a way out of a potential tight spot for the UK government, whose commitment to the EU Code and publicly announced tightening of the licensing regime for sales to Israel would have made the granting of a licence for the direct export of such equipment to Israel problematic. It also warns that the arms industry is likely to take 'this new environment into consideration with regard to future organisational structure' (Saferworld 2003a, 8-9); that is, the loophole is likely to grow. Saferworld points out that the US government takes a more restrictive approach to onward sales, insisting on authorisation prior to re-export and applying its veto rights where it deems necessary. Saferworld 'recommends that the UK does the same, and moreover encourages other EU member states to do so as well' (2003a, 9). CAAT also commented on the case, arguing that the UK government 'has chosen against upsetting defence business ties with America at the expense of betraying its own rules on arms exports'. CAAT's concern about the new guidelines is that 'the economic interests of UK arms exporters will always override humanitarian, economic and security concerns in the final recipient country' (CAAT 2002b).

The effect of the change in policy brutally came to light in 2009, following a UK government investigation into whether UK-supplied equipment was used by the Israeli Defence Forces in Operation Cast Lead, its attack on Gaza in December 2008/January 2009. Foreign Secretary David Miliband stated that 'Many of the licences we have identified covering military equipment were for components for incorporation into US-manufactured platforms which were then re-exported to Israel' (Miliband 2009). While the government claims that no licences have been granted for the export of F-16 components directly to Israel since 2002, 'British made components for F–16s have been exported to the United States where Israel was the ultimate end-user' (Miliband 2009). This admission that UK-produced components have ended up in Israel via the US came despite Jack Straw stating that the new guidelines included consideration of 'the export control policies and effectiveness of the export control system of the incorporating country' (Straw 2002). The UK government's position is that the US licensing system is 'strong and effective', and where there are concerns over Israeli behaviour, there is dialogue between the US and Israel (Straw 2002). This is despite the fact that there would be criticism in the UK (from NGOs, activists and likely the Parliamentary Committees on Arms Export Controls) if F–16 components were exported directly to Israel. According to Straw, the defence relationship with the US is 'fundamental to the UK's national security as well as to our ability to play a strong and effective role in the world' and 'There are also wider benefits to the UK's national security of maintaining a strong indigenous defence industrial capability.' Meanwhile, there remains a need for 'a break to the cycle of violence' in Israel/Palestine (Straw 2002). This example demonstrates that, while the US has probably the world's strictest licensing regime on paper, the political interpretation of it is driven by foreign policy concerns. It also demonstrates UK subordination to the US: while components for F–16s are unlikely to be licensed directly to Israel from the UK (although this is not impossible), the desire to maintain transatlantic defence-industrial relations – even in instances in which US policy is at odds with publicly stated UK policy – takes precedence over human rights commitments.

How are we to understand NGO activity in relation to transatlantic military production and trade? Whether insider or outsider, NGOs' main concern is with the risks to export control standards. Saferworld and BASIC operate from within the government's parameters, making technical suggestions on how best to promote both UK defence-industrial concerns and export licensing standards, trying to encourage the UK government to make the most restrictive decisions within the parameters of this debate. This is partly an issue of strategy: defence collaboration is a complicated, technical issue that does not arouse as much public opposition as arms exports to human rights abusers. However, it is also emblematic of the depoliticisation of high-technology intra-Northern trade in which defence collaboration and integration are seen primarily as an issue of economics (Spear and Cooper 2006, 316). This is in turn linked to the changing nature of Northern militarism, in which increased war preparation has corresponded to a partial demilitarisation of society (Shaw 1991, 13).

While it makes strategic sense for NGOs to remain on depoliticised ground to try to generate policy change, they simultaneously reproduce the wider structures of Northern militarism. Thus, if we understand defence collaboration as a particular mode of imperial relations – the integration of the UK into a US-led system of military production and trade, in which the US exercises greater control over the direction and content of that system – then we can see the NGOs as facilitating imperial relations while attempting to maintain maximum room for manoeuvre for the UK as subordinate party. The distinction between the NGOs is that, as seen previously, BASIC also has a more critical position on the overall orientation of UK defence and foreign policy than does Saferworld. While it engages in technical, problem-solving action in this instance, it also attempts to change the overall parameters of debate. CAAT, meanwhile, does not engage on the government's terms: while its critique may be transformist, there is less chance of its proposals being taken seriously.

European defence collaboration

NGOs raise similar concerns regarding European defence collaboration as they do in relation to the transatlantic relationship. In 2000, the governments of France, Germany, Italy, Spain, Sweden and the UK signed a Framework Agreement to restructure their defence industries in order to 'create the political and legal framework necessary to facilitate industrial restructuring in order to promote a more competitive and robust European defence technological and industrial base in the global defence market and thus to contribute to the construction of a common European security and defence policy' (Secretary of State for Foreign and Commonwealth Affairs 2001). NGOs raised a series of concerns around export controls, highlighting the risk that the harmonisation of European defence-industrial policy could involve a downward shift to the lowest common denominator, undermining the EU Code and privileging industry priorities over human rights (BASIC 2001b; CAAT 2000; Miller and Brooks 2001; Saferworld 2000).

The arguments of BASIC and Saferworld in relation to the harmonisation of European arms trade processes are based on the recognition that such moves are necessary as US predominance in the arms industry becomes more pronounced. European governments need to 'sustain a viable defense industry capable of competing in the global market, while reducing the barriers to trade between allies' and to address the 'ever-growing costs of today's weaponry' (Miller and Brooks 2001). The Framework Agreement thus presents European states with a valuable opportunity to consolidate their defence industries and coordinate armaments policies, in which the main concerns are to make sure loopholes are closed and transparency is increased (Miller and Hitchens 2000). Similarly, Saferworld argues that the Framework Agreement is 'one of the most significant manifestations of the efforts of European governments to facilitate the restructuring of the European defence industry in the face of contracting markets and stiff competition from the US' (Saferworld 2004b, 41). In Saferworld's view, 'the rationalisation of the EU defence industry, with increasing emphasis on collaborative projects, is not in itself a problem,' but 'it is important that national parliaments and the public have confidence that European defence collaboration is not being sustained by means of a less rigorous application of the EU Code of Conduct (or the Consolidated Criteria in the UK)' (Saferworld 2004b, 41).

BASIC and Saferworld thus seek to generate a win–win situation in which Europe makes its industrial base more competitive, which contributes to European security and yet does not undermine arms export controls. Saferworld and a coalition of European NGOs have mapped out a set of practices they believe would mitigate these risks. This includes the use of 'no re-export without permission' clauses, implementing delivery verification procedures for all non-EU destinations, reserving the right to conduct post-export end-use monitoring, and implementing end-use control on all EU-licensed production overseas (EU NGOs 2008).

BASIC has sought to maintain European independence of the US without undermining NATO, and to reduce the emphasis on military power and militarised responses to security problems. For example, it has tried to wend a path between accepting that 'Europeans do indeed need to take on more responsibility for international and regional security' and challenging the idea that this should result in 'an increase in defence spending and the purchase of US military equipment' (BASIC 1999). It emphasises that there are 'important transatlantic differences between perceptions of future threats and how to respond to them', even if the EU's focus on terrorism, WMD proliferation and failed states, and organised crime suggest that 'threat perceptions are converging' (BASIC 2006a). In the past, BASIC has tried to promote the creation of 'civilian intervention units', which would help the EU 'get rid of its image as a military "dwarf", avoid large increases in defence spending, simultaneously put the brakes on US unilateralism, and attack the real threats to its security' (BASIC 1999; also BASIC 2000). Within this, the UK should play its 'familiar British role of intermediary between the United States and the European Union', in which it can 'carve out a middle ground between American desire to preserve a strong NATO and European aspirations for greater independence in the realm of European security' (BASIC 2001c). BASIC has thus historically tried to prevent the rise of a militarised EU, tried to maintain European independence from the US, and emphasised soft or non-military security issues and responses.

For CAAT, developments in the EU raise similar concerns to those in play domestically, namely the relation between arms capital and the state. For example, it is concerned that the European Constitution includes a 'very definite commitment to develop European military capacity' (CAAT no date, g), which received 'little public or parliamentary debate' (CAAT 2005c). Discussions have 'been dominated by the arms companies' and there has been 'little consideration to the many non-military ways Europe might become more secure and have a positive influence on the world' (CAAT 2005c). This is echoed in an ENAAT report, which argues that the arms industry has 'been successful in finding itself a niche in the common security and defence policy. Lobbying by the industry has impacted directly on political decisions,' with the result that the European Constitution 'points towards a more aggressive way of solving conflicts, a less restrictive arms export policy and an increase in violence and armed conflict' (Broek and de Vries 2006, 4).

European developments neatly demonstrate the differences between the NGOs. BASIC and Saferworld both make technical, problem-solving interventions oriented towards progressive goals around transparency and rigorous control. They accept US military predominance and European states' concerns to sustain market share within this, whilst not letting the Europeans forget the importance of arms control. But BASIC does this in the context of a wider critical intervention over the orientation of European security. CAAT, meanwhile, criticises the privileged role enjoyed by arms companies and its effect on European security policy. Generally, however, NGOs do not work extensively on transformations within Europe as an issue in and of itself; their greater concern is with the implications for human rights, development and conflict of exports from Europe to the South.

Significant tranches of intra-Northern military production and trade are simply not an issue for NGOs, therefore. For example, there has been little public commentary or advocacy work in relation to the establishment of the European Defence Agency in 2004, which aims to increase cooperation between EU member states on defence-industrial policy, strengthening the European defence technological and industrial base, ensuring security of supply between EU member states, and creating a competitive Europe-wide defence equipment market (Business and Enterprise Committee *et al.* 2008). The signing of a European Code of Conduct on Defence Procurement in 2006 means that EU state governments will not automatically purchase military equipment from national suppliers, but will purchase from each other if the offer is the best available (EDA, cited in Hartley 2008, 304). This process of rationalisation is aimed at closing the EU–US gap in defence R&D spending (Hartley 2008, 205). While NGOs were central to the development of the code of conduct on arms exports, they have paid little attention to this code of conduct, which rationalises procurement practices within Europe. Again, this is an indication of the depoliticisation of militarism within the European world: it is significant that military production is seen mainly as a defence-industrial issue rather than a security issue.

Conclusion

NGO activity in relation to intra-Northern military production and trade is noticeable by its relative absence compared to the issues of North–South relations and small arms proliferation. Of the seven NGOs, only BASIC, CAAT and Saferworld work on issues around domestic procurement and exports to NATO allies, and for none is this strand of work the main focus. Saferworld allies a reformist argument with an insider strategy, trying to improve processes within the parameters of the government's position. BASIC combines a transformist argument with an insider strategy, combining problem-solving technical suggestions with a more critical vision. And CAAT allies a transformist argument with an outsider strategy, but focuses more on the implications for exports to the South than on Northern forms of militarisation.

NGOs' work on challenging mainstream economic arguments around the arms trade has served to disturb deeply entrenched understandings of the issue. In the case of DESO, they succeeded in shutting down one of the arms trade's institutional pillars of support, although the significance of this change remains to be seen. However, the NGOs work with different ultimate goals in mind: CAAT wants to abolish the arms trade, BASIC and Saferworld want it to be better regulated. On arms sales to NATO governments, BASIC and Saferworld concentrate largely on improving the processes of defence collaboration, trying to minimise risks to export controls and generate a win–win situation in which defence collaboration is deepened but peace and security are maintained. BASIC combines this with a wider argument more in line with CAAT's overall philosophy, challenging the broader parameters of international security. CAAT's campaign mission includes progressive demilitarisation within the UK, but in practical terms it does not comment on domestic procurement or military posture to any extent. It is the only NGO to focus on the relationship between arms capital and the state, but does little to place this in an international context. CAAT has links to ENAAT, some of whose members have produced campaign materials on transformations in the arms trade in other European countries and at the EU level, but it does not campaign on intra-Northern military production and trade as an issue itself. Like some other European anti-arms-trade groups, CAAT lacks the resources even to monitor, let alone campaign on these developments in any sustained way. In addition, its focus on state–capital relations sidelines wider questions of militarism that cannot be reduced to the socio-economic power of arms companies, as discussed in Chapter Two.

Broadly, NGOs have tried to reintroduce North–South security concerns into defence technology debates, to move them away from being purely an industrial or trade issue, and so compelling the government to consider arguments it would rather ignore. But the nature of the security concerns they raise are all about lowest common denominator export controls and the risk of exports to the South, rather than about the militarisation of the North, and the role of the North in creating conflict (using the weapons it produces and trades) and in promoting hegemonic forms of militarisation. The representation of international relations produced by NGO activity thus sidelines Northern militarism as part of the problem of organised violence in world politics. The role of military production and trade in facilitating Northern military hegemony and its disproportionate role in creating a technology- and capital-intensive world military order are not raised. But these are important in themselves as an issue of peace, security and sustainable development, and are a significant part of the picture when thinking about the context behind Southern militarisation.

In terms of strategy, Saferworld and BASIC start from the government's current position and from within this framework try to argue for the most progressive course of action. They work within the government's position but try to push it in the direction of tighter export controls and more rigorous processes. BASIC combines this with a more transformist vision of European and transatlantic security. In part, this is due to BASIC's specifically transatlantic focus – it claims to be the first and only one of its kind and engages in elite-level advocacy strategies, which allows it to promote the incremental steps necessary for its vision to be operationalised. CAAT's starting point, in contrast, is much further outside the parameters of the government and industry. Its transformist vision shares some characteristics with BASIC's, but goes further as it is based on abolition of the arms trade. The key difference between the NGOs is the access that they have, based in part on the depth of their research, their expertise on certain technical issues, as well as their perceived respectability.

Overall, BASIC's and CAAT's strategies are largely complementary. Saferworld's shift towards the Control Arms agenda suggests a cleavage in NGO work, however. That is, the Control Arms campaign may undermine more transformist change because it does not require the world's largest military spenders and exporters to alter the orientation of their practices. These omissions are playing out in the Arms Trade Treaty. As will be discussed in the next chapter, NGOs such as Amnesty, Oxfam and Saferworld have been highly active and influential on issues such as the effects of armed violence on individuals and communities, the impact of conflict on security and development, and the opportunity costs of military spending, in which resources are diverted away from social needs towards military concerns. But all of this needs to be done without impinging on states' domestic concerns, as the Arms Trade Treaty is a UN process in which state sovereignty must be paramount. One key effect of this framing of the problems posed by the arms trade, combined with the relative inattention to intra-Northern production and trade, is that transfers within NATO, the creation of a hegemonic military culture and rise of global military spending (dominated by the US) are off the agenda. In this way, the 'problem' of the arms trade is seen as internal to the South and the practical mechanisms being promoted by NGO activism serve to subject Southern states to a degree of scrutiny that the US and European states are spared.

5 • Disciplining the South: Development and Human Rights Concerns in the Arms Trade

Arms sales to the South are widely understood to pose a risk to human rights and development. Amnesty International, Oxfam, Saferworld and CAAT have been the main UK-based NGOs active on these issues, working to document the impact of the arms trade and tighten national and international controls on transfers. The former three have worked individually and collaboratively through the UK Working Group and, since 2003, the Control Arms campaign. 2003 signalled a shift in their strategy: in the late 1990s and early part of the 2000s, they focused on controversial UK arms transfers, criticising the government for contravening its publicly stated commitments to promote development and human rights. With the advent of the Control Arms campaign, they moved towards a more collaborative relationship with the UK government in order to promote the idea of an Arms Trade Treaty internationally and have been strategically effective in terms of garnering state and industry support for their efforts. They have researched and developed detailed arguments about the impact of arms transfers on human rights and development, devising methodologies for assessing this impact and minimising harm. As mentioned earlier, CAAT, meanwhile, has retained its outsider strategy of criticising controversial UK arms transfers that damage human rights and development in the South, focusing in particular on the relationship between arms capital and the UK state.

NGOs are agreed that arms sales to the South pose a risk to human rights and development. Where they differ is in their identification of the source of the problem and in their proposed solutions. For Amnesty, Oxfam and Saferworld, the problem is inadequate regulations and enforcement, and the task is to help states – first the UK, and later internationally – understand the risks better, devise tighter mechanisms for controlling arms exports, and develop dialogue with importers about what transfers are appropriate. For CAAT, the problem lies in the pro-export orientation of major arms producers, in particular the UK, and the close relationship between arms companies and the government. CAAT makes tougher demands through more confrontational means, and while it makes policy demands, it is less focused on engagement in the detail of the policy process. Despite this difference, there is a key shared effect of NGO activity, namely the production of the South as a site of intervention and its resultant disciplining. This takes place in different ways: the Control Arms NGOs represent the problems of human rights and development to be internal to the South, and their solutions obscure the international dynamics of a world military order. While CAAT focuses its attention on the role of the UK in creating human rights and development problems in the South, its relative inattention to demilitarisation within the North means that its call for an end to arms sales to the South would, if successful, freeze a profoundly hierarchical status quo.

'Genuine defence needs': the impact of the arms trade on development

In 1997, as the EU Code of Conduct on arms sales was being negotiated, Oxfam called for it to prohibit 'arms sales to countries where military spending is far beyond what can be justified by the country's genuine defence needs' (Oxfam 1997). It welcomed developments under the Labour government such as the Mauritius mandate, whereby 'export credits for poor highly-indebted countries will only support productive expenditure' (Oxfam 1998). Oxfam proclaimed this decision not to underwrite the sale of military weapons to 62 of the world's poorest countries a 'turning point' (Oxfam 2000) and called on the UK government to 'lead by example towards an international agreement which places a presumption against subsidies for the export of arms to all Least Developed Countries [LDCs] which are in violent conflict' (Oxfam 1998).

Concerns over the impact of arms transfers on sustainable development were aggravated by the UK government's decision in 2001 to grant a licence to BAE Systems for the export of a £28m air traffic control system to Tanzania, despite the opposition of DfID, the Treasury, the World Bank and the International Civil Aviation Organization (ICAO). This case is a turning point in NGO activity in relation to development and the arms trade (see Stavrianakis 2005). NGOs, working individually and as part of the UK Working Group coalition, argued that the case signalled an inconsistency in the UK government's approach to development. That is, the export to one of the world's poorest countries of a technically doubtful set of equipment that was bad value for money undermined the government's wider commitments to development. For Saferworld, arms exports are the 'missing link' in an otherwise progressive foreign policy agenda: '[t]he government's wider objectives – on human rights, conflict prevention and sustainable development – are being undermined by a failure to effectively control arms exports' (Mepham, quoted in Saferworld 2002b). And Oxfam described the Tanzania decision as revealing an absence of 'joined-up government' (Watkins 2001) as the DTI (as it was then known) was able to issue the export licence without consulting DfID or considering the implications for poverty in the recipient country. The controversy over the Tanzania case led to an inter-departmental review of development concerns in arms export licensing within the UK government. This led to procedural rather than policy changes, including the strengthening of DfID's methodology for assessing the impact of arms transfers on development (see Hewitt 2002). Overall, as noted in Chapter Three, the case is indicative of the tensions between two networks of actors, with DfID, the Treasury, the World Bank and the ICAO losing out to the combined strength of the DTI, MoD, BAE Systems and Prime Minister Tony Blair.

Up until the early 2000s, NGOs such as Oxfam and Saferworld were critical of the UK government for contravening its publicly stated commitments with regard to development concerns. They issued public criticism, helping generate public controversy over particular cases such as that of Tanzania, using them as examples to push the government towards stricter controls on arms exports. The Tanzania case reinforced the view within the NGO world that the way to improve export controls is to pursue scandal. For Oxfam campaigners, dialogue will not in and of itself create change – the government needs to feel it is under pressure. Scandals such as this one are understood as 'real policy shifters'. For Oxfam, part of the problem was that 'the mystical "other factors"', such as the government's relationship with BAE Systems, overrode the commitment to sustainable development. But there is also a wider problem, in that governments neither know what they mean by sustainable development, nor have benchmarks for it, nor know how to implement it, according to one campaigner. NGOs thus used public controversy to reinforce their work in trying to support the hand of institutionally weaker elements of the state, in particular DfID. The publication of DfID's assessment methodology gave NGOs an insight into how the government dealt with development concerns in arms export licensing, and they set to work on trying to improve these processes so that better decisions could be made in future. This led to a significant tranche of work, linked to the Control Arms campaign, aiming to promote best practice in terms of development concerns in arms transfers.

The Control Arms NGOs raise a number of concerns about the impact of the arms trade on sustainable development, which, following the 1987 Brundtland Report, they define as 'a combination of economic growth and social progress that meets the needs of the present without compromising the ability of future generations to meet their own needs' (Control Arms 2004a, 9). The basic problem is that 'excessive or inappropriate arms purchases are a drain on social and economic resources which developing countries cannot afford' (Control Arms 2004a, 3; also Chanaa 2005). Such purchases increase the risks to developing countries of debt, corruption and waste (Oxfam 2008a, 10). More specifically, irresponsible arms transfers drive up defence spending in developing countries, and impose opportunity costs through the diversion of resources away from education, healthcare, and social development (Oxfam 2008a, 2). The misuse of arms has negative social effects in terms of conflict, gender inequality, as well as denial of healthcare, education, livelihoods and aid (Control Arms 2003, 31-3; Control Arms 2004a, 18). And the increased risk of corruption and wasteful expenditure associated with arms transfers cost developing countries more and can lead to purchases that are neither part of a national security strategy nor cost effective (Oxfam 2008a, 2; Oxfam 2008b, 2). For the world's poorest states, irresponsible transfers threaten progress towards realisation of the Millennium Development Goals, a set of eight internationally agreed goals to be met by 2015, including ending poverty and hunger (Control Arms 2004a, 3).

Oxfam accepts that 'developing countries may need to import arms to meet legitimate self-defence and security needs', but believes 'spending *beyond those legitimate needs* represents a waste of resources that are often crucially needed for social development' (Oxfam 2008a, 10, emphasis in original). Governments 'often fail to allocate the necessary resources to prioritise the full realisation of [economic, social and cultural] rights in favour of other areas, such as military expenditure' (Oxfam 2008a, 10). As such, 'developing countries generally spend a *greater proportion* of their national product on arms than do rich countries' (Control Arms 2004a, 18, emphasis in original). In particular, 'Anti-democratic, highly militarised governments are more likely to expend resources on the military at the expense of development spending' (Control Arms 2004a, 30). Overall, while the UN Charter recognises states' right to self-defence, the Control Arms NGOs point out that it:

also applies principles of sustainable development to the use of arms, calling for the 'establishment and maintenance of international peace and security with the least diversion for armaments of the world's human and economic resources'. Yet with global military spending amounting to $839bn a year, the combination of 'over-armament and under-development', to quote a phrase first coined two decades ago, is still a real problem. (Control Arms 2003, 9)

States thus have an obligation to balance defence and security against sustainable development (Oxfam 2008b, 5).

This argument is shared by the UK government. Citing research it commissioned from the Bradford University Centre for International Cooperation and Security, DfID acknowledges that 'few developing countries have their own indigenous arms industries, so they are often dependent on arms imports'. It draws a distinction between 'responsible' transfers, which 'can create space for development by helping governments provide security for their populations', and 'irresponsible' transfers, in which 'the costs of maintaining and using these weapons...can divert resources from development spending on areas such as education or health' (DfID 2007b, Ev. 75-6). The attention paid to development concerns in arms export licensing is an attempt to protect against such negative impacts.

As part of these efforts, DfID takes into account the cumulative impact of all imports to the recipient country, not just those from the UK (Hewitt 2002), as 'The sustainable development impact of a proposed export can only be assessed in the context of the purchasing country's military procurement policy as a whole' (Defence Committee et al. 2003, 31). More widely, the UK government also played a pivotal role in promoting a new approach to military expenditure internationally in the 1990s. Known as the 'process' or 'governance' approach, this focuses on reform of the processes by which military spending occurs in the South, with a view to creating the stability necessary for development (Omitoogun 2003, 262-7). This replaced an approach in which aid donors imposed upper limits on military expenditure as a condition for receiving aid, and was part of the wider rise of the agenda of promoting good governance and democracy in Northern foreign policy.

CAAT agreed with Amnesty, Oxfam and Saferworld that the Tanzania decision was a contravention of the government's publicly stated commitments to sustainable development, arguing that '[i]t is senseless of the UK government to preach about debt reduction and development when on the other hand it is enticing countries to purchase high-tech weaponry they can ill afford' (CAAT 2001). It also agrees that 'arms transfers and militarism are the triggering factors that exacerbate debt, poverty and conflict' and that there is a 'lack of coherence in the Labour government's current policies towards debt, development and the arms trade' (Willett 1999). The arms trade 'saps the resources of already-poor countries' and is a diversion of resources away from social spending (CAAT no date, h).

While there is broad agreement within the NGO world that the arms trade threatens development in the South, CAAT focuses on a different reason for the emergence of this problem. It argues that global poverty is 'the direct result of the structure of the global economy, of the history of colonialism and of political decisions being made today' and that the arms trade creates as well as exacerbates poverty (CAAT no date, h). This is in contrast to Oxfam, for example, which acknowledges that 'responsible, regulated transfers of military and security equipment can assist a state to fulfil its legitimate defence, military, and policing needs, which can help to provide the security and stability necessary for development' (2008a, 4). For Oxfam and the other Control Arms NGOs, the issue is that governments do not know how best to operationalise development concerns, while for CAAT, the problem is that international structures are weighted against the South. CAAT also directs its attention towards the North, arguing that, while developing countries are being forced to eradicate subsidies as part of a free trade agenda, arms companies receive substantial levels of corporate welfare (CAAT no date, h; also CAAT 2005d, 5). Governmental support for arms exports – in the UK and elsewhere in the G8, which make up the world's largest arms exporters – 'is one of the major obstacles to the eradication of poverty' (CAAT 2005d). So for CAAT, the use of public money to subsidise UK arms exports is directly linked to the continuation of global poverty. While the UK is unusual in having development concerns represented within the licensing process, CAAT argues that DfID has had 'a negligible impact', especially in light of the wider support for arms exports from the rest of the government (CAAT 2005d). Thus, while the NGOs are agreed about the risks to development of arms transfers, CAAT makes a wider structural argument and pays more attention to the role of the North in creating the problem.

A key theme of NGO concerns around the impact of the arms trade on development has been the pivotal and devastating role of corruption. For NGOs such as those involved in the Control Arms campaign, the wider context is one of good governance. According to them, 'There is a clear relationship between governance standards and military spending' (Control Arms 2004a, 65). They argue that 'context is critical'. That is, 'understanding the relationship between governance and arms imports' should be central to any assessment of the impact of arms transfers on development (Control Arms 2004a, 46). As well as macroeconomic considerations, prior to a licence being approved, there should be 'investigation of defence procurement and budgeting practices...the extent of involvement of a wider range of actors in the decision-making process, and the degree to which efforts are made to assess any potential impact upon sustainable development' (Control Arms 2004a, 46).

Corruption in the international arms trade is a core concern for CAAT. It agrees with the broad thrust of the other NGOs' position, namely that it 'can undermine democratic accountability and divert resources away from healthcare or education into projects which are amenable to bribery' (CAAT no date, i). CAAT places more emphasis than the other NGOs on the role of corruption in the arms trade, however, citing the work of political economist Joe Roeber, who argues that 'Corruption is not peripheral; it acts at the centre of procurement decision-making.' The arms trade is thus 'hard wired for corruption' (Roeber 2005, 7). As well as the impact on the South, CAAT emphasises that 'Western arms companies routinely bribe the political and military elite of countries into buying arms they may not need and certainly cannot afford' (CAAT 2002c). For example, in relation to Saudi Arabia and based on files found in the National Archives, CAAT argues that corruption, in the form of bribery, has been central to UK–Saudi arms deals, that the UK government has been aware of this, and that bribes have taken the form of payments by BAE Systems, approved by the MoD, to bank accounts under the personal control of Prince Bandar (CAAT 2009c; CAAT 2006c). And in the Tanzania case, CAAT echoed Clare Short's scepticism that the deal could have been made cleanly (CAAT 2002c).

This scepticism seemed well founded, given that the Serious Fraud Office (SFO) launched an investigation in July 2004 into allegations of bribery against BAE Systems in relation to arms deals with Tanzania, as well as Chile, the Czech Republic, Qatar, Romania, Saudi Arabia and South Africa. The Saudi investigation was dropped in 2006 under political pressure, and CAAT and The Corner House's legal challenge against the government failed to reinstate it. The other cases remained ongoing for a further three years. By late 2009, the SFO had indicated its intention to prosecute BAE in relation to Tanzania, South Africa, Romania and the Czech Republic (BBC 2009a). However, in February 2010 a plea bargain settlement was reached in which BAE agreed to pay fines of US$400m (£255.7m) to the US Department of Justice and £30m to the SFO to settle corruption charges. It pleaded guilty to charges of conspiring to make false statements to the US government in relation to arms sales to Saudi Arabia, the Czech Republic and Hungary, and of failing to keep reasonably accurate accounting records in relation to sales to Tanzania. The payment to the SFO is described in news reports as an ex-gratia payment for the benefit of the people of Tanzania (Press Association 2010a, 2010b).

In the Tanzania case, press reports state that middleman Sailesh Vithlani, who grew up in Tanzania but holds a British passport, had been 'closely working with a network of influential local partners, including high-ranking government officials both current and retired, to secure the lucrative contracts' for this and other controversial procurement deals, and admitted that BAE Systems secretly paid massive commissions into a Swiss bank account (Leigh 2007; ThisDay 2007). This was done via two parallel arrangements: a conventional commission agreement of 1 per cent, a standard figure in arms deals; and 'a second, more unusual agreement', in which an offshore company owned by BAE, Red Diamond, paid a further $12m into a bank account in Switzerland under the personal control of Vithlani (Leigh 2007). This amount represented 30 per cent of the contract price, which, 'coupled with the use of a Swiss bank account and apparent double sets of agency agreements, would normally arouse suspicions of possible bribery', according to investigators (Leigh 2007).

NGO suggestions for change
The Control Arms NGOs attempt to promote a more progressive policy in order to protect development concerns over the arms trade. They point out that states already have obligations under existing international, regional and national control regimes. These require that arms transfers must not seriously undermine a state's economy, hinder or obstruct sustainable development, or involve excessive or unjustifiable diversion of resources from social to military expenditure. Further, they must involve the least diversion of human and economic resources for armaments, include a role for democratic institutions to determine defence policies, and be compatible with the technical and economic capacity of the recipient state (Oxfam 2008b, 5). On this basis, NGOs argue that states are already bound to consider the developmental impact of arms transfers when making import and export decisions: the challenge is to get states to take this responsibility seriously. NGOs thus emphasise that governance, appropriateness, affordability, and economic, industrial, technological and technical capacity need to be taken into account (Control Arms 2004a, 5).

NGOs want 'policy makers to think imaginatively about how to engage with importer governments to change their spending priorities in line with the country's most pressing development needs' (Control Arms 2004a, 30). As such, they have produced reports to 'assist states to apply sustainable development standards when making decisions regarding international arms transfers through the application of a clear and consistent procedure' (Oxfam 2009, 1) and have developed an intricate methodology that could help states assess development concerns (see Control Arms 2004a; Oxfam 2009). For example, they measure a state's level of development by looking at its Human Development Index (HDI) value. They differentiate three thresholds: if a state's value falls below 0.65 (i.e. for 'countries of low development'), they then ask if this is a financially significant transfer. For those states with a value between 0.65 and 0.85, they ask if the transfer is of a scale that might have an impact even in countries with higher levels of development. For those states with a value above 0.85, 'no further analysis on sustainable development is required' (Control Arms 2004a, 56-7). There follows a complex and nuanced methodology for assessing the impact of transfers through the use of indicators and triggers, as a contribution to improved policy making. This activity is part of an insider strategy of strengthening the hand of more progressive, but institutionally weaker elements of the state.

In terms of suggestions for change, CAAT does not engage in the micro-detail of policy. It focuses instead on what it understands as the political and socio-economic impetus to the overall policy of support for the arms trade, which has a problematic effect on development. So, for example, it calls for an end to the use of public money to subsidise arms exports when arguing about the impact of the arms trade on global poverty (CAAT no date, h). It exposes what it sees as the roots of the problem, criticising for example state-sponsored marketing, export credits, military aid and support for arms fairs as institutional pillars of support for the arms trade (CAAT 2005d). Thus the policy changes that it calls for are less directly focused on the practicalities of assessing sustainable development and aim to remove the factors that it understands to create the problem in the first place.

Assessing NGO activity

How are we to assess NGO activity in relation to development concerns in the arms trade? An initial point to note is that governmental commitment to development is, at times, more rhetorical than real. The Tanzania case, in which a licence was granted despite the express opposition of DfID, demonstrates the institutional weakness of DfID compared to the MoD,

DTI and the Prime Minister. As we will see in Chapter Six, DfID has considerable leverage on issues where there are no significant defence-industrial interests at stake, such as on small arms and conflict. However, when DfID makes demands that challenge the interests of arms companies and the elements of the state allied with them, it loses out. Development concerns thus do not carry as much weight within government as defence-industrial concerns. As a result, the NGOs are all agreed that one positive step would be for DfID's voice to be heard more strongly within the institutional process. They adopt different strategies in relation to this, with Oxfam and Saferworld working to strengthen DfID's hand, making policy suggestions that would improve these processes, while CAAT stands outside of the intricacies of the policy process, demanding an end to the close relationship between arms companies and the government and the associated policy choices that stem from this. But they are agreed that DfID should have a greater say on arms export issues, and are agreed that expensive arms sales to poor countries are a contradiction in an otherwise benevolent development agenda.

The development agenda itself deserves scrutiny, however, its relative weakness within government notwithstanding. The general principles articulated in the development agenda, shared by NGOs, DfID and research centres associated with them, are only deemed to be an issue in the South. That is, NGOs and donor governments only articulate military spending and associated arms transfers as a problem if a country is poor. The UK government, for example, states that its development guidelines are designed to 'pick up high value applications to the poorest countries' (DfID 2007b, Ev. 74); a University of Bradford report as part of the DfID-funded Armed Violence and Poverty Initiative focuses on arms transfers to those parts of the world 'where the bulk of the world's poor people are, that is in "developing countries"' (Bourne *et al.* 2004); and the three-tier set of HDI thresholds articulated by NGOs serves to focus attention on poor countries. In a similar fashion to the Saferworld *Audit* methodology discussed in Chapter Four, NGOs' use of particular categories for deciding which international arms transfers to focus on excludes the North from scrutiny by definitional fiat. CAAT also reproduces this focus on the South, to a degree. Although it focuses on the role of the North in promoting arms sales and 'considers that high military spending is unacceptable and only reinforces a militaristic approach to problems', one of its priorities is to end arms sales to 'countries whose social welfare is threatened by military spending' (CAAT no date, a) − but not to countries whose military spending is simply high in either absolute or relative terms. Arms transfers within the North and certain regions of the South thus fall out of the equation as an issue of political economy. In terms of the world's largest importing states and regions, India is included in the development matrix because of its low HDI figure (see Control Arms 2004a; Oxfam 2008a), while China rarely is; the Middle East is effectively excluded by definitional fiat, while Europe definitely is.

Because they are not poor, the military budgets of the US (as the world's leading spender on the military and the world's largest arms exporter) and other major Northern military spenders are not deemed problematic. NGOs point out that the UN Charter calls for the least possible diversion of expenditure towards armaments, and that global military spending is enormous. They acknowledge that even in more advanced economies, 'there is no consensus that increased military expenditure is good for the economy' (Control Arms 2003, 36). But in practice, the North is excised from view. In a report devoted to the impact of the arms trade on development, for example, the US military budget does not get a mention (Control Arms 2004a). And the North is excluded from analysis of development concerns in arms export decisions through the three-part threshold, discussed above. Because 'the economic and financial burden of US military spending (i.e. its share of GDP and of total US government outlays) is lower now than during previous peak spending years in the post–World War II period', US military spending is not often considered to be part of the problem, despite being higher in 2007 than at any time since World War Two and accounting for 45 per cent of global arms spending in 2007 (SIPRI 2008). So NGO practice disconnects the problems of development and the arms trade from the wider context of global military spending.

As a result of this definitional exclusion, a whole international apparatus is mobilised to scrutinise and improve Southern behaviour, as seen in the Control Arms NGOs' suggestions for change. The practical focus of NGO activity is on encouraging Southern states to make better decisions. This means that Southern states' domestic procurement practices are open for scrutiny while arms producers' practices are not. This is despite NGOs' direct rebuttal of the argument that 'Any treaty that focuses on transfers of weapons and not on production discriminates against states that do not manufacture weapons' (Saferworld 2009a, 6). Countering this, they argue that the Arms Trade Treaty would not interfere with the international arms trade and military procurement that allows states to satisfy their needs for self-defence and law enforcement, 'providing there is not a substantial risk that the arms internationally traded and procured would be used for serious violations of international law or would contribute to excessive or destabilising accumulations of arms' (Saferworld 2009a, 6). But again, the exclusion of production within the major arms-producing countries and the assumption that 'excessive' procurement is only a problem in the South sidelines the North and North–South relations from view and focuses attention on the South.

The issue of corruption provides a useful illustration of the politics of North–South relations in the arms trade. Clare Short suggested that bribery had been involved in the Tanzania deal; this was deemed problematic as corruption would undermine the promotion of good governance and the promotion of development measures. The Serious Fraud Office's commitment to this investigation was reaffirmed around the same time that its inquiry into BAE Systems's relationship with Saudi Arabia was dropped under political pressure. The intervention of the Prime Minister in this judicial matter amounts to political interference in the rule of law, and means that BAE Systems is effectively above the law; these are precisely the issues addressed in the good governance agenda that Tanzania – but, it seems, not Saudi Arabia or the United Kingdom itself – is subject to.

CAAT and The Corner House attribute difficulties in prosecuting corruption to the lobbying power of multinationals like BAE Systems. Countering the UK government's claim that it is committed to tackling corruption and bribery, and challenging the OECD judgement that the UK lacks the political will to enforce anti-bribery rules, they argue, rather, that the UK government 'has devoted immense political will over the years to the task of protecting a few powerful companies from prosecution for bribery' (Corner House 2008). They point to the direct involvement of Prime Minister Tony Blair in calling off the Serious Fraud Office investigation, sending a personal minute to the Attorney General detailing his concern over the 'critical difficulty presented to the negotiations over the Typhoon contract' (Corner House and CAAT 2007). While all remaining SFO inquiries were also eventually dropped, the fact that they lasted longer than the Saudi inquiry suggests that corruption may sometimes be challenged under the good governance agenda, but only when it is not strategically risky to do so and, even then, this remains extremely difficult. CAAT's work in challenging the harder Saudi case head on thus signals a transgressive role, with the result that it was infiltrated by private intelligence companies working for BAE Systems, as discussed in Chapter Three. After the announcement of the BAE plea bargain, the High Court granted an injunction preventing the director of the SFO from taking any further steps in relation to the settlement, pending its decision as to whether or not to grant CAAT and The Corner House permission to apply for judicial review of it. Their application was refused in late March 2010; the two NGOs eventually decided not to appeal and withdrew their application in April 2010, citing a misleading statement from the SFO as giving cause for concern over the UK's commitment to tackling corruption (CAAT 2010a; CAAT 2010b).

The Saudi example signals the ambivalent position of the Middle East in the development agenda and the wider issue of the different ways in which the South becomes a focus of concern. The Middle East is one of the world's largest arms-importing regions and demonstrates the importance of a North–South understanding of the US-dominated world military order. During the Cold War the Middle East accounted for approximately one-third of the world's arms deliveries, dropping to one-quarter in the 1990s and one-fifth in the early 21st century. In 2007, seven of the ten states in the world with the highest military burden (share of military spending in GDP) were Middle Eastern (Oman, Saudi Arabia, Israel, Jordan, Lebanon, Yemen and Syria) (figures in this paragraph from Perlo-Freeman 2009b). The US is far and away the largest supplier to the region, accounting for 53 per cent of the volume of deliveries in 2004-8, compared to 46 per cent in the previous five years. The next largest suppliers in 2004-8 were France (16 per cent), Germany (8 per cent) and Russia (7 per cent). Russia's main customers in the region are Iran (59 per cent of Russian deliveries to the region), Iraq (16 per cent), Egypt (15 per cent) and Syria (10 per cent). Few states in the region have significant arms industries – only Israel is able to produce a wide range of platforms, and it remains dependent on the US, which supplies nearly 99 per cent of Israel's conventional weapons imports (Perlo-Freeman 2009b).

The Middle East's levels of military spending relative to spending on health and education are often high but oil wealth means its states are not widely understood as poor. In terms of the Human Development Indicators used by NGOs to operationalise their methodology, Middle Eastern states are variously categorised as middle- and high-income countries. NGOs do argue that the high levels of military spending is a problem in the region: they point out that the region 'spends an average of $12bn per year on arms imports; more than Latin America, Africa, and Asia put together' (Control Arms 2004a, 12) and is well represented in the list of states that have high levels of military spending relative to spending on health and/ or education (Control Arms 2004a, 8). But there is a slippage when the development agenda is operationalised. That is, while Middle Eastern states are included as part of the problem of high military spending, when NGO methodologies are put into practice they fall out of the equation because their HDI values are too high. So there is an internal differentiation within the category of the South, with the Middle East excluded to some degree from the NGO-DfID interventionist apparatus. Within the South, the focus is selective, directed towards the world's poorest states; but they are not the biggest purchasers of military equipment. The Middle East, in contrast, one of the world's largest arms-importing regions, is heavily supported in this by the arms capital–state network of suppliers. As Robinson (1996) argues, liberal, polyarchic social relations will be promoted where possible, but authoritarian ones where necessary.

Overall, rather than signalling a contradiction, arms sales can be understood as the other side of the coin to the provision of aid, in that they are both oriented towards stability within the global economy. The development agenda itself is framed by neoliberal capitalist principles, claiming to tackle poverty but functioning to entrench it further in many parts of the world. The primary effect of DfID's policies is to open up new markets for private capital (Burnell 1998, 792); this is in line with the policies of the major multilateral donors, the IMF and World Bank (Cammack 2004). Privatisation and liberalisation are a reason for the persistence of poverty in large parts of the world rather than a solution to it as state services are sold off to private companies, leaving local populations to pay for services or go without. The aid agenda has its counterpart in the arms sales agenda, through which the institutional and material basis of many states is shaped. Military purchases may well signal the squandering of precious resources by profligate elites. But they are also activities sanctioned and conditioned by a global military culture that valorises capital-intensive forms of militarisation, by the dominance of militarised representations of status on the world stage, and by state formation in the South under conditions of hierarchy.

Human rights, arms trade wrongs

Human rights have been a key element of the post-Cold War international security agenda, alongside poverty reduction and good governance. In the UK, the Labour government entered power in May 1997 promising to 'put human rights at the heart of our foreign policy' (Cook 1997a) and to 'refuse to supply the equipment and weapons with which regimes deny the demands of their peoples for human rights' (Cook 1997b). NGOs had worked hard with Labour, both in opposition and as it came into government, to get these promises made and institutionalised through regimes such as the EU Code. As with development concerns, there has been a shift over time.

Once Labour came to power, NGOs such as Amnesty, Saferworld and CAAT issued public criticism of exports that contravened the government's publicly stated commitments to human rights. Around 2003, with the launch of the Control Arms campaign, Amnesty and Saferworld moved towards a strategy of working with the UK government to try to promote human rights concerns internationally. CAAT, meanwhile, continues its strategy of criticising the government, based on an understanding that UK policy has not become more benevolent and that national controls are the best way forward.

Arms sales to Indonesia are a good example of the different arguments and strategies in play in the various NGOs. In the late 1990s and early 2000s, the common ground between Amnesty, CAAT and Saferworld was the view that arms exports to Indonesia ran the risk of being used in internal repression and were thus often problematic in terms of the UK government's export licensing rules. The three NGOs all situated UK arms exports in the wider context of a repressive Indonesian military. In 1997 Amnesty argued that the Indonesian armed forces 'are focused primarily on combating internal dissent rather than external threats' and 'play a prominent role in quelling peaceful dissent and, in their handling of violent dissent...have frequently committed serious human rights violations' (Amnesty International 1997). In 2000 Amnesty argued that, without end-use monitoring of transfers, the UK arms export policy towards Indonesia 'risks making a complete mockery of the government's manifesto commitment to a human rights-centred foreign policy' (Amnesty International UK 2000). Amnesty used UK arms exports to Indonesia as an example of how 'G8 member states are undermining their commitments to poverty reduction, stability and human rights with irresponsible arms exports to some of the world's poorest and most conflict-ridden countries' (Amnesty International UK 2005b). Saferworld made a similar argument, asking 'whether the government's human rights foreign policy is being undermined by increased arms sales to countries with dubious human rights records' (Saferworld 2003b). It argued that: '[c]hanging the approach to arms exports does not require a seismic shift. It is a natural conclusion of the government's existing policies on human rights and development' (Saferworld 2002c).

CAAT agreed with Amnesty and Saferworld that arms sales to Indonesia contravened the UK government's arms export guidelines. However, it emphasised the continuities in British policy between Conservative and Labour governments, and understood this as part of an effort to maintain the Indonesian state's territorial integrity and suppress internal dissent in a state rich in natural resources in which Western powers such as the United

Kingdom have heavily invested (Gilby 1999). It argued that Labour made rhetorical commitments to human rights but continued to promote arms trade to Indonesia, 'breaking almost all its "ethical" guidelines' (Gilby 2001). Although the Suharto regime fell in 1998 and Indonesia now has electoral democracy, the underlying necessity for Indonesia's elite to purchase foreign arms remains intact, as the vested interests of Indonesian and Western elites remain in place, according to one researcher. This claim is the basis of CAAT's understanding of why exports to Indonesia continue to be licensed. That is, the Indonesian military is 'doing a job for Britain' in making sure 'nothing too radical' happens, such as Acehnese independence, economic autonomy and nationalisation of its oil companies. In this sense and in this researcher's opinion, the military 'keeps a lid' on the situation in Indonesia and 'keeps the capitalist wheel ticking [sic]' despite recent moves towards political liberalisation.

NGO calls for action in response to arms sales to Indonesia are illustrative of their different orientations and strategies. In 1997 Amnesty called for 'a halt to the transfer of a range of military and security equipment and training to Indonesia, including armoured personnel carriers, assault rifles and sub-machine guns, and lethal training for the special forces' because of human rights violations that occurred during the quelling of dissent in East Timor, the impunity that surrounded this, and the 'high potential for misuse' of transferred equipment (Amnesty International 1997). In 1999 it called for 'an immediate moratorium on the sale and supply of military equipment and training to Indonesia that could be used to commit human rights violations' in East Timor (Amnesty International UK 1999). And in 2000 it argued that the EU should not lift its arms embargo on Indonesia 'for now', and that the EU 'must not resume the sale of arms or security equipment likely to be used to commit human rights violations in Indonesia' (Amnesty International 2000). However, three days after Amnesty made this call, the embargo was lifted.

Amnesty International is not opposed to arms sales *per se* and only opposes the transfer of military, security and police equipment and other items 'where such transfers can reasonably be assumed to contribute to human rights violations within AI's mandate', as one International Secretariat researcher described it. According to a UK section campaigner, Amnesty needs to be able to link specific types of equipment to particular human rights violations in order to be able to protest against their use. This has always been a core feature of Amnesty's work on arms issues, and tallies with its wider approach to campaigning, which stipulates that information must be credible and impartial. Amnesty thus 'does not call for "comprehensive" arms embargoes unless it can make a reasonable assumption that "all" the arms likely to be transferred will be used for serious human rights violations', according to one campaigner. The East Timor crisis highlighted the tensions in such an approach. Amnesty faced the difficulty of obtaining reliable evidence of the use of Hawk jets in East Timor, much of which only emerged two or three years after the atrocities. It was difficult for Amnesty to win over UK parliamentarians; as one researcher articulated the problem, 'you have to "reasonably demonstrate" that it might be used – that's a political reality NGOs have to deal with. Just because we assert it doesn't mean the government should believe it.'

Saferworld does not have a stated policy of impartiality on arms questions in the way Amnesty does, yet in practice it works in a similar manner, matching types of equipment to violations and calling for restrictions on arms exports tightly matched to the level of violations. In this, both NGOs mirror the UK government's case-by-case approach to arms export licensing, but try to encourage a more restrictive interpretation. In 1999 Saferworld called for an immediate EU arms embargo on Indonesia unless it agreed to the deployment of an international peacekeeping force in East Timor. Saferworld's audits of the government's 2000 and 2001 Annual Reports state that the NGO would have expected a full embargo to have been in place in relation to exports to Indonesia in these years (Saferworld 2002a, 87; Saferworld 2003a, 91). In contrast, in 2003 and the first half of 2004, it would have expected a 'selective embargo' to have been in place (Saferworld 2005, 51). In its analyses of UK export policy as a whole, Saferworld seldom calls for a full embargo, which is 'a broader brush'; this is not its *modus operandi*. Rather, it focuses on specific transfers and on what is happening at a particular time, according to one staff member. Like Amnesty, Saferworld would only expect an embargo to be in place in instances where all transfers of military equipment run the risk of being used in human rights violations.

In contrast, since CAAT's formation in the mid 1970s it has been calling for a full UK and international embargo on military exports to Indonesia (CAAT no date, j). It calls for a more severe response than Amnesty and Saferworld, based on the nature of its understanding of the function of arms exports and because of its outsider strategy. Its stance towards government is more confrontational than that of Amnesty and Saferworld; it argues that the UK government knows that UK-supplied military equipment has been used in Indonesia, and that extra-judicial killings occurred, yet, according to Labour, 'because no one can prove that UK equipment actually killed people, there is apparently no risk that Alvis equipment [armoured vehicles] might be used for internal repression!' (Gilby 2001, emphasis in original). In 2003, along with 89 other signatories and led by Tapol, an Indonesian human rights campaign group, CAAT renewed its call for an international arms embargo on Indonesia, covering military, security and police equipment and with retrospective application (Tapol 2003). And in the late 1990s, it launched (and duly lost) a case of judicial review against the UK government, arguing that it had contravened its publicly stated commitments by licensing arms exports to Indonesia. In line with their overall respective strategies, CAAT is able to call for an embargo and apply for judicial review in a way that Amnesty and Saferworld are not.

NGO's activity in relation to Indonesia is indicative of their national focus in the late 1990s and early 2000s. While CAAT has continued with this approach, Amnesty and Saferworld started to internationalise their policy proposals with the advent of the Control Arms campaign in 2003. This was partly because UK government practice was understood to have improved, and partly a response to the globalisation of the arms trade and to the international nature of arms supplies to regions and states in which human rights are threatened. The Control Arms NGOs make a variety of demand- and supply-side arguments. On the supply side, the Control Arms NGOs emphasise the role of major arms suppliers, pointing out that the permanent members of the UN Security Council (China, France, Russia, the UK and the US) account for 88 per cent of the world's conventional arms exports, 'and these exports contribute regularly to gross abuses of human rights' (Control Arms 2003, 5; also Control Arms 2005, 3). Amnesty International, in particular, is critical of supplier governments whose 'irresponsible and poorly regulated' arms transfers contribute to human rights violations in countries as diverse as Colombia, Côte d'Ivoire, Guatemala, Guinea, Iraq, Myanmar, Somalia, Sudan and Uganda (Amnesty International 2008). Overall, the Control Arms NGOs conceptualise the problem of human rights as one of 'inadequate controls and poor practice in implementing and enforcing those laws and regulations which do exist' (Control Arms 2005, 4).

In addition, NGOs point out the negative turn under the War on Terror, in which 'some suppliers have relaxed their controls in order to arm new-found allies against "terrorism", irrespective of their disregard for international human rights and humanitarian law' (Control Arms 2003, 4). Some of these recipients of military aid 'are armed forces which have committed grave violations of human rights and have been identified in the State Department's own human rights report as having a "poor" human rights record, or worse', such as Armenia, Azerbaijan, Afghanistan, Colombia, Georgia, Israel, Nepal, Tajikistan, Turkey and Yemen. According to the NGOs, 'Arms and military assistance are being offered as a geopolitical inducement, with few, if any, conditions to protect human rights' (Control Arms 2003, 41-2).

In terms of the demand side and the role of arms transfers in human rights violations in the South, the Control Arms NGOs point to a pattern of state forces abusing their authorised weapons during peacetime. This is a result of inadequate wages, impunity for armed extortion and corruption, limited or non-existent training, and insufficient attention being paid to internationally agreed rules for law enforcement in law, regulation and training (Control Arms 2003, 18). The NGOs argue that 'In many countries the resources for policy, equipment and training are insufficient' and in some cases there is a 'deliberately repressive government policy'. Taken together, these conditions mean that 'police resort to excessive and arbitrary force, or use firearms for unlawful killings and as an instrument of torture and ill-treatment against suspects' (Control Arms 2004b, 2). In such circumstances, international arms transfers 'send a message that the behaviour of such groups is tolerated, even supported, by the international community' and 'may actually encourage further atrocities by reinforcing the impunity with which they operate' (Control Arms 2003, 19). This undermines internationally agreed standards for the use of force: UN standards 'require government law enforcement agencies to avoid using force when policing unlawful but non-violent assemblies, and, when dispersing violent assemblies, to use force only to the minimum extent necessary' (Amnesty International 2008, 37; also Control Arms 2004b, 17). The key underlying question is 'what constitutes legitimate force?' That is: 'Police must sometimes be permitted to use force or lethal force...But the force used must not be arbitrary; it must be proportionate, necessary and lawful' (Control Arms 2004b, 2).

NGO suggestions for change

In order to promote the observance of human rights in the arms trade, the Control Arms NGOs call for an Arms Trade Treaty that would codify states' existing obligations under international law and institutionalise the distinction between legal and illegal transfers. For Amnesty, while 'it is clear that many states recognise that their obligations under international human rights law have application to transfers of conventional arms, the rigorous and consistent application of these obligations must also be prioritised' (Amnesty International 2008, 109), and the Control Arms campaign is an attempt to generate shared standards and implementation mechanisms for doing this.

A key element of this is the distinction between a 'preventive' and a 'punitive' approach, with Amnesty and the other NGOs promoting the former (Amnesty International 2008, 115). A preventive approach aims to 'prevent arms transfers where there is credible and reliable information indicating there is a substantial risk that a particular group, such as the security forces, will use those arms for serious violations or abuses of human rights', while a punitive approach 'reduces the decision-making process to one where states that are seen to have unspecified "bad human rights records" cannot receive any transfers of arms' (Amnesty International 2008, 115). Operationalising a preventive approach involves a three-part process, in which a proposed recipient's attitude towards and respect for international human rights law is assessed, the nature of the equipment and stated end-use are taken into account, and a decision is made as to whether there would be a 'substantial risk' of misuse (Amnesty International 2008, 116-22).

CAAT's wider position on human rights is similar to its strategy towards Indonesia, in that one of its priorities is to 'end exports to oppressive regimes' (CAAT no date, a). It thus exemplifies what the Control Arms NGOs would criticise as a punitive approach, as it has called for embargoes on arms transfers to states such as Indonesia and Israel, and has engaged in or supported judicial review applications to challenge government policy towards such states. More widely, its work on challenging the role of DESO in promoting arms exports, and its criticism of the close relationship between arms companies and the UK government are aimed at tackling the impetus to such controversial exports.

Assessing NGO activity

The Labour government strengthened the institutional position of human rights concerns by increasing the size of the Human Rights Policy Department by one-third on coming into power in 1997 (Foley and Starmer 1998, 472). However, it remains weak compared to pro-export forces. There is also a significant degree of continuity between Labour and Conservative policy, the rhetoric of the 'ethical dimension' to foreign policy notwithstanding. For example, the Conservative government claimed that 'All applications to export defence equipment are carefully scrutinised on a case-by-case basis', that the UK government 'would not grant an export licence if we thought that the equipment was likely to be used for purposes of repression', and that 'We do not allow the export of arms and equipment likely to be used for repressive purposes against civil populations' (cited in Phythian 2000a, 159, 151, 160).

NGOs point out that arms suppliers' commitments to human rights are often more rhetorical than real. The NGOs have spoken out against the way in which political allegiances and necessities have overridden human rights commitments under the War on Terror, for example. However, the human rights agenda itself needs to be subjected to scrutiny. As expressed and operationalised by NGOs, it ultimately directs our attention to the South and to particular instances of violations, rather than to wider and more enduring North–South relationships. NGOs acknowledge that some states have policies of repression towards (parts of) their populations, but they do not articulate the North–South relations that foster these, or lend a rationale to such repression.

Human rights violations cannot adequately be understood outside the social, political and economic context in which they occur. As Robinson argues, structural violence (across the South but also in parts of the North) generates collective protest, which is met by the state with repression, which transforms 'structural violence into direct violence'. This means that '[t]he structural violence of the socioeconomic system and violations of human rights are different moments of the same social relations of domination' (Robinson 1999, 62). Understanding human rights violations in this way requires us to focus on the systematic oppression that is a necessary feature of capitalism, sometimes supported by patrons through arms supplies, rather than the violation of individual (mainly civil and political) rights, and to understand that '[r]epressive practices are a means to an end, the end being the maintenance of some form of political power' (Desmond 1983, 40, 76, 129).[1]

Having said this, the Control Arms NGOs' preference for a preventive rather than a punitive approach could be oriented towards such a position, but it would require the application of political judgement to do so. The alternative, punitive approach or, in CAAT's language, a call for embargoes and an end to exports to Southern states involved in undesirable practices, has its own problems. Imposing blanket bans on exports to the South without an equal focus on military production within the North would freeze a profoundly unequal hierarchy in access to the means of violence, and would further reproduce a binary opposition between a South that is not to be trusted with weaponry and a North that is, disconnecting the South from the wider military order in which particular types of military equipment and violent conflict are points on a spectrum.

Conclusion

The impact of the arms trade on human rights and development is a core concern for Amnesty International, Oxfam, Saferworld and CAAT. The former three, working in concert, first as part of the UK Working Group on Arms and later the Control Arms campaign, enact a reformist argument through an insider strategy that complements research and the provision of methodologies and attempts to bolster more progressive elements of the state with public pressure through the mass membership of Amnesty and Oxfam. CAAT, meanwhile, has a more transformist argument that focuses on the institutional pillars of support for controversial arms transfers, and an outsider strategy that criticises rather than engages in micro-level policy work.

Despite this difference, their arguments reproduce the South as a site of concern and perpetuate practices that discipline the South. The threat to development and human rights posed by the arms trade is deemed to be internal to the South and disconnected from the world military order and capitalist state system of which it is a part. NGOs' language that Southern states often 'fail' to allocate adequate resources or to pay attention to human rights makes these practices sound like an aberration. But arms imports are an element of the process of building and reinforcing the material and institutional basis of the state, in which relations with external patrons are crucial. While CAAT is more critical of Northern involvement, its relative inattention to Northern levels of military production and trade mean that it reproduces an understanding that arms transfers are only an issue in relation to the South.

To make this argument is not to deny that direct physical repression and the opportunity costs of high military spending are more keenly felt by the populations of the South. In this sense, the goal of NGO activity – to focus on where the world's problems are worst, to make a difference where it is most needed – stems from a progressive impulse. However, the failure of NGOs to make the connections between Northern military preponderance, the world military order, dependent state formation and the very real problems faced by populations in the South may ultimately be a disservice to their cause.

Note

1   Desmond's argument is particularly interesting as he is a former head of the UK section of Amnesty International, and his book *Persecution East and West* is a biting critique of the organisation.

6 • Disarming the South: Small Arms and Conflict

Small arms emerged as an issue on the international security agenda in the 1990s and since then significant efforts have been made by states, inter-governmental organisations (IGOs) and NGOs to combat the spread and misuse of what have been labelled 'the real weapons of mass destruction' (IRIN 2006, 3). One of the dominant themes of international action on small arms is conflict prevention: the spread of small arms is widely understood to hamper development, which is said to threaten security and make conflict more likely. This has been a key post-Cold War security theme, in both its human security and, since 9/11, anti-terrorism variants. The UK state and UK-based NGOs – in particular Saferworld, IANSA, International Alert, Amnesty International and Oxfam – are significant actors on small arms issues, pushing for stronger international controls on the supply of small arms and also funding and implementing numerous projects in the South.

To a significant extent, there is a shared international understanding among states, IGOs, research centres and NGOs of the role of small arms in conflict. UK-based NGOs such as Amnesty International, BASIC, International Alert and Saferworld have contributed significant intellectual and practical resources to this common understanding and the practices associated with it. These practices contribute to complementary attempts to shore up state monopolies on violence and control over weaponry as part of tighter supply-side regulation, and to remove weapons from societies and improve governance processes as part of a demand-side agenda. NGOs play a role in buttressing, and often significantly extending, efforts to contain conflict in the South, transform Southern societies, and maintain coercive state capacities. In relation to small arms, then, NGOs are integrated into state and intergovernmental practices to a significant degree, much more so than on issues in the wider arms trade.

However, NGOs have had limited influence on more sensitive issues for states, such as civilian possession and transfers to non-state actors, and in regions such as the Middle East, and South and East Asia, in which small arms debates remain tied to a state security framework of understanding. In addition, their activity is challenged by more conservative NGOs, in particular the National Rifle Association (NRA) in the US. Small arms control is a battleground in more ways than one; there is contestation within what we normally call global civil society as to what counts as progressive action.

The conflict–security–development nexus

International NGOs, predominantly those based in the US and Europe, have been key participants in international action on small arms. There was an 'explosion of interest' in small arms in the mid 1990s (Krause 2001, 11) with scholars such as Michael Klare and Edward J. Laurance, and NGOs such as Human Rights Watch, BASIC and International Alert all active in researching the effects of small arms proliferation, some supported financially by grants from foundations such as the Ford Foundation, the MacArthur Foundation and the Ploughshares Fund (Garcia 2006; Krause 2001). There are competing views on whether NGOs were agenda-setters or followed a trend started by states. Some analysts point to the activity of NGOs in the early 1990s, prior to UN Secretary General Boutros-Ghali's 1995 call for micro-disarmament, as evidence of their agenda-setting role (Anders 2005, 179; Lumpe *et al.* 2000, 7; Joseph and Susiluoto 2002, 129, 132). Others argue that NGOs were slow to react to his call, as in 1992 there were only two NGOs addressing small arms issues, 12 in January 1998, but over 200 by February 2000 (Laurance and Stohl 2002, 4, 27). Further, some analysts argue that 'state sponsorship was critical to the success of NGO initiatives' (Cattaneo and Krause 2004, 20) and 'NGO activities *followed* rather than preceded state or intergovernmental efforts' (Krause 2001, 29, emphasis in original). In such a view, small arms 'was first a scholar-led subject, then became state-led, and then NGOs followed the states' (Garcia 2006, 24, 40).

What is clear is that a small number of NGOs raised concerns about small arms in the early to mid 1990s, amongst them the UK-based organisations discussed here. While Amnesty International had attempted to launch a project and campaign on small arms as far back as 1979, its efforts did not gain traction at the time (Krause 2001, 11). In the 1990s, NGO efforts were more successful, not least because of Amnesty International's and Human Rights Watch's documentation of the role of small arms in the Rwandan genocide, a key 'shock' event that spurred international action (Garcia 2006, 93-4; Amnesty International 1995; Human Rights Watch 1995). International Alert, for example, claims to have been working on small arms issues since 1994, 'when we identified unregulated small arms proliferation and misuse as one of the world's most pressing security issues' (International Alert 2006a). BASIC's 'Project on Light Weapons' consisted of approximately 200 organisations and individuals around the world at its height in the late 1990s, with the project's membership directory serving as 'an invaluable tool in jump-starting the IANSA network' (BASIC 2001d), which is now one of the leaders of the Control Arms campaign. IANSA was established in 1998 off the back of the success of the landmines campaign, with Amnesty International, BASIC, International Alert, Oxfam and Saferworld as key actors from the outset (see Clegg 1999; Garcia 2006; Grillot et al. 2006; Krause 2001 for overviews of the formation of IANSA).

As this chapter demonstrates, the issue of the degree of NGO influence over states is somewhat misplaced. More significant is the shared agenda between the NGO small arms control community and states such as the UK, Canada, Norway, Switzerland and Belgium, and IGOs such as the United Nations. The activity of this network needs to be assessed against other networks, such as that between pro-gun organisations, in particular the National Rifle Association, and the US state, as well as the structurally more powerful network that links arms capital to states.

NGOs, research centres, IGOs and governments broadly share a common understanding of several of the issues at stake, including the links between conflict, security and development, the significance of the illicit trade, and the importance of holistic solutions. The discourse linking conflict, (in)security and (under)development is central to NGOs' arguments concerning small arms. In 1996, BASIC claimed that 'the issue at hand is frighteningly obvious: light weapons kill, maim and destroy; they cause instability and prolong wars; they promote a culture of violence which is gaining momentum around the world; and they divert much-needed resources away from social and economic development' (Dyer and Goldring 1996). Saferworld pithily summarises the threat posed by small arms: 'The proliferation and misuse of small arms and light weapons fuels crime, exacerbates violent conflict and undermines development' (Saferworld no date, d). According to IANSA, small arms proliferation is 'a serious humanitarian challenge with implications for development human rights and global justice' (IANSA no date, b). Amnesty International and Oxfam emphasise 'the abuse of arms which fuels conflict, poverty, and violations of human rights'; whilst this applies to all categories of weapons, 'small arms have a particular role to play in contributing to poverty and suffering' (Control Arms 2003, 4, 19). The effects of the misuse of arms 'increase poverty and derail development' in the long run as part of a 'vicious circle' of arms abuse in which, citing Paul Collier's work on the relationship between conflict and development, 'Poverty fuels conflict...[and] conflict fuels poverty' (Control Arms 2003, 35, 24). International Alert describes small arms as 'a barrier to peace', in that they 'fuel conflict...and are used indiscriminately to kill, injure and intimidate civilians' (International Alert 2006a). CAAT does very little work on small arms specifically, but argues that they 'contribute to the initiation of violent conflict' and are 'instrumental in perpetuating it'; their widespread availability 'can erode negotiated peace settlements, hampering conflict resolution and post-conflict reconstruction' (CAAT no date, k). For CAAT, 'the unrestrained flow of arms into...unstable states has fanned the flames of conflict and undermined development'; Africa, in particular, is 'awash with small arms' (O'Grady 1999).

Within this broadly shared understanding amongst the NGO community, organisations contribute their respective expertise in human rights, development, conflict prevention and so on. There is a practical difference in emphasis between those focusing more on the security and supply side, and those on the community safety and demand side. However, there has also been a shift in the overall orientation of the NGO world over time, expanding from a narrow to a broader view of security, with a related shift from a focus on arms exports to human security. This shift, and the different emphases within it, chimes with wider developmental responses to small arms that have taken root since the late 1990s. Overall, they are an attempt to recognise the spectrum of armed conflict, organised crime and social violence in the South that blurs the boundaries between war and peace (Small Arms Survey 2003, 125-7; also Geneva Declaration 2008; Muggah and Krause 2009).

The security-oriented, supply-side approach can be seen in the work of Amnesty and Saferworld, which criticise state practices of small arms exports that contravene international standards. For example, Amnesty is highly critical of states that have supplied small arms (as well as other weaponry) to the Sudanese government, which has committed 'serious violations of international human rights and humanitarian law', through both its own security forces and the government-aligned Janjawid militias.

Small arms have been transferred to Sudan from China, France, Saudi Arabia, Switzerland, Ukraine and Brazil (brokered via the UK), with arms also being smuggled in from southern Sudan, Chad, Libya, and central Africa; Human Rights Watch has identified arms from Belgium, Hungary, Israel, Russia, South Africa, Sweden, the UK, the US and the former Yugoslavia (Amnesty International 2004, 26-32; see also Small Arms Survey 2007). In Iraq, the widespread distribution of arms under Saddam Hussein's regime, combined with the failure of US-led forces to prevent human rights abuses or control the circulation of weaponry (including US Department of Defense-sponsored supplies to the Iraqi security forces), have contributed to the 'massive proliferation and misuse of weapons' and 'grave human rights violations and abuses' being suffered by the civilian population. Part of the problem has been 'irregularities in the supply by US private contractors of arms to Iraq', including deliveries of small arms and ammunition from Bosnia by a foreign airline company under a Department of Defense contract, which then could not be accounted for by US officials when the shipment arrived in Iraq (Amnesty International 2008, 39-49).

Saferworld criticises the UK government for small arms transfers that contribute to conflict, in the same vein as it criticises other controversial exports. For example, Saferworld would have assumed that a 'presumption of denial' would have been in operation in 2003–4 with regard to arms exports to Morocco, given the violent political conflict over control of Western Sahara. Thus, the UK government's licensing of over 300 small arms to the Moroccan government was of concern to Saferworld (Saferworld 2005, 51-2). It has also begun assessing Russian and Chinese involvement in the arms trade more closely. It argues that Russian small arms transfers to Ethiopia and Eritrea prior to and almost immediately after the one-year UN embargo announced in May 2000, as well as transfers to Syria, which the US and Israel suspect of being diverted to Hezbollah forces in Lebanon, raise concerns that Russian export controls are not in line with international standards and that the Russian government 'has had few qualms about selling arms to any state that seeks to purchase them' (Holtom 2007, 28-33). Since 2006, Saferworld has also been working to engage Chinese academics and think tanks on arms trade and control issues (Saferworld 2009b): the Control Arms campaign points to the transfer of Chinese weapons to Zimbabwe in 2008 as another example of the need for shared international standards (Control Arms 2008b).

Controversial transfers that contravene international standards would be covered by the NGOs' vision of an Arms Trade Treaty. The Treaty would stamp out illicit transfers and codify states' responsibilities to abide by human rights and international humanitarian law in the licensing process. It would apply as much to the US and European states as to Russia, China and other second-, third- and fourth-tier military producers. As well as this supply-side work, within the NGO community there is also an extensive demand-side focus on community safety. The NGOs are agreed on the conflict–security–development arguments, but there is a division of labour between them in terms of their focus and, relatedly, their attitude to government. Amnesty International can issue no-holds barred criticisms of state-sanctioned arms supplies, for example, while Saferworld issues more 'constructive' messages and IANSA and International Alert focus on the community level.

The work of IANSA and International Alert is emblematic of the community safety agenda, which they both argue has emerged from the field in the South. IANSA's campaigning, grounded in 'the direct experience of our members on the frontline' as well as in research and information, is based on the idea that most gun deaths 'occur in the developing world' and gun injuries 'are especially disabling in the developing world' (Peters and LaPierre 2004). Sexual violence, domestic violence and the incidence of child soldiers are all exacerbated by small arms, which also obstruct peacekeeping, and hinder development, investment and tourism (Peters and LaPierre 2004). Driven by the demands of its Southern members, IANSA has focused on action on gun violence as a community safety issue, according to a member of the secretariat. It pays particular attention to the gendered nature and public health implications of gun violence (IANSA no date, c). The secretariat avoids the term 'small arms', which it argues places the issue in a security framework; rather, it talks in terms of 'guns'. Over the years, IANSA has shifted away from the technical side of small arms control towards gun violence, crime and gun control, according to one member of the secretariat.

International Alert was heavily involved in the late 1990s and early 2000s in research and advocacy on small arms issues, through the Biting the Bullet project and the build-up to the 2001 UN conference on small arms. Biting the Bullet was a collaboration between International Alert, Saferworld and, in its earlier days BASIC, which aimed to 'contribute to a better-informed debate on small arms issues' in preparation for the 2001 UN conference (International Alert 2006c). More recently, BASIC has left the group and Bradford University's Centre for International Cooperation and Security has come on board. As described by one of its staff members, International

Alert has now moved away from small arms as an issue in and of itself towards broader security sector reform (SSR) and disarmament, demobilisation and reintegration (DDR) work, of which small arms remains a part, but as an element in a much broader and holistic approach. Much of the work by the NGO small arms community in the late 1990s and early 2000s was pitched at the level of the UN and global policy, which had little traction with International Alert's field programmes. As a result, according to one International Alert staffer, there was a strategic refocusing of the organisation and a shift towards broader peacebuilding and community safety work. The organisation's focus on issues such as gender came out of its field experience and, more broadly, it saw its role as linking up security and development. On issues such as SSR, DDR and small arms, practitioners have tended to go in with a very technical, military focus; International Alert sees its role as trying to make these reform processes more holistic.

There is a somewhat uneasy relationship between small arms control and the wider movement for conventional arms control (see also Goldring 2006). There is a lack of clarity about the relationship of the UN small arms control process that started in 2001 (discussed in more detail below) and the Arms Trade Treaty process that began two years later. On one hand, most Southern IANSA members think the Arms Trade Treaty is about small arms, according to a secretariat staffer, and most government ministries involved in the process tend to be ministries of internal or foreign affairs, rather than defence. On the other hand, the demand side of the Control Arms campaign has not been developed as fully as the supply side, according to an Amnesty campaigner. Community safety issues are too disparate, requiring tailored solutions, to be developed into a global campaign. This tension has been present since the inception of the Control Arms campaign. As described by an Amnesty campaigner, there are those in the South who argue that international arms controls are nothing to do with why armed violence is a problem; what is needed is work on conflict resolution and demand-side issues. In this view, the North may well be responsible for the supply of weapons but tighter controls would not necessarily stop the suffering caused by armed violence. Overall, of the four main frames for understanding small arms – human rights/humanitarian, conflict/disarmament, development, and faith – the former two are predominant in Northern NGOs and the latter in the South, which also has a greater emphasis on local gun control initiatives (Krause 2001, 19).

Despite this difference in strategic focus, there is a broadly shared understanding within the NGO community of the threat posed by small arms.

They are seen as key contributors to conflict and poverty, undermining development and facilitating human rights abuses. Notably, NGOs – predominantly International Alert, Oxfam and Saferworld – have participated in promoting a developmental frame of reference for understanding small arms issues. They have worked in tandem with DfID to try to naturalise this understanding, recommending, for example, a change in language from 'small arms proliferation' to 'armed violence reduction' (DfID 2003). Working with DfID and academics based at the University of Bradford's Centre for International Cooperation and Security, NGOs have played a significant role in making the development aspects of small arms issues more explicit, as have research centres such as the Small Arms Survey (e.g. Small Arms Survey 2003, ch. 4). There is thus extensive co-production of the narrative of conflict, security and development between government, the non-governmental sector and academic centres.

With the advent of the War on Terror, there has been an additional narrative of anti-terrorism in relation to small arms, operating alongside the conflict–security–development argument. While the conflict–security–development nexus is most actively promoted by Sub-Saharan African and European states, states such as the US, Russia, Israel and India articulate a mainly anti-terrorist framing. Conflict, security and development, and anti-terrorism are politically liberal and conservative, respectively, but share some discursive features and have some similar practical effects. Both narratives are mainly articulated in relation to the illicit trade in small arms, for example, and the proliferation of weaponry in areas of conflict, weak governance and poverty is understood as simultaneously a conflict prevention, development and anti-terrorism issue. The political salience of the argument that poverty is a significant cause of terrorism means that small arms programmes are understood not just as a benevolent act but also as in Western states' interests as well (Hewitt 2003; Thomas 2006). In this way the small arms agenda is emblematic of the merging of security and development, which 'now qualifies as an accepted truth of the post-Cold War era' (Duffield 2005, 142).

A key feature of international action on small arms, in both its conflict–security–development and anti-terrorism guises, is the emphasis on the illicit trade. The use of small arms by state militaries and the importance of civilian possession in some states, particularly the US, have meant that a condition of the emergence of any degree of international consensus has been the definition of the problem as that of the illicit trade. As Garcia argues, while the destruction and disposal of surplus weapons, regulation of illicit arms brokering, and marking, tracing and record-keeping are becoming shared international norms, the regulation of civilian gun ownership and prohibition on sales to non-state actors, as well as transparency in small arms transfers, are weak or failed norms (Garcia 2006).

International action on small arms control occurs largely under the umbrella of the UN Programme of Action to Prevent, Combat, and Eradicate the Illicit Trade in Small Arms and Light Weapons, in All Its Aspects (hereafter, Programme of Action), agreed at the 2001 UN Conference on the Illicit Traffic in Small Arms and Light Weapons in All Its Aspects (hereafter, UN Conference). The parameters of debate at the Conference were limited by states such as the US, with its insistence that the conference should only tackle the illicit transfer of military-style weapons, keeping both the legal trade and civilian possession off the agenda. The US also vetoed any discussion of restrictions on transfers of small arms to non-state actors. States such as China, Cuba and several Arab states also stymied the conference, but the US was the most vocal and inflexible of critics (Stohl 2001; Krause 2002, 247). While states such as Iran, Syria and Russia were also opposed to regulating transfers to non-state actors, they 'stood back and let the United States be blamed' (Garcia 2009, 156). As Karp puts it, 'Although it is unfair to tar the United States with all the blame for the failure of the conference, it could have done more than any other country to rescue it' (Karp 2002, 188). It was only after a number of African states backed down over the need to regulate civilian possession and transfers to non-state actors that agreement on the Programme of Action was reached (Stohl 2001).

Despite the lack of normative consensus around the problem posed by small arms and serious divisions between states on scope and remit, particularly in relation to the regulation of civilian possession and transfers to non-state actors, participants agreed that the illicit trade in small arms and light weapons was a serious international security threat that required some form of concerted action (Karp 2002; Krause 2002; Stohl 2001). The Programme of Action states that the 'illicit trade in small arms and light weapons in all its aspects sustains conflicts, exacerbates violence, contributes to the displacement of civilians, undermines respect for international humanitarian law, impedes the provision of humanitarian assistance to victims of armed conflict and fuels crime and terrorism', as well as being closely linked to 'terrorism, organised crime, trafficking in drugs and precious minerals' (United Nations 2001). This narrative is simultaneously wide-ranging in scope, given the variety of threats with which small arms are associated, and narrow – it is the illicit trade in small arms that is deemed to be the core problem. NGOs have had to accept powerful states' redlines on this issue, but have used avenues such as the Biting the Bullet project to try to push for a more inclusive approach that would start to address the legal trade as well, or at least address the links between the licit and illicit trades.

NGOs emphasise the importance of holistic solutions. For example, BASIC describes small arms as 'symptoms of other problems, such as disputes over resources and borders. To reduce the killing, we must understand and overcome many obstacles. While some are political and military in nature, others are economic, social, and even psychological' (Dyer and Goldring 1996). A resource pack produced jointly by International Alert, Oxfam and Saferworld argues that 'tackling the demand for small arms covers a broad range of issues, including structural and deep-rooted problems such as poverty, inequality, bad governance, and underdevelopment, in addition to more specific measures to tackle the weapons themselves' (Coe and Smith 2003, 21). The Biting the Bullet project emphasises the importance of linking small arms measures to security sector and governance reform programmes, arguing that 'In post-conflict situations, action on small arms is crucial to supporting governance and building security' (Biting the Bullet 2006, 231). And the Control Arms campaign emphasises the need for both supply-side and demand-side measures, using the analogy of mopping up the floor and turning off the tap (Control Arms 2003, 5). This is deemed a progressive impulse, a move away from state-based security towards human security that also contributes to wider international security by removing the causes of conflict.

This holistic approach is echoed by the UK government, which presents its commitments on small arms as part of its 'wider conflict prevention, poverty reduction and defence diplomacy policy and programmes' (FCO 2006). There is a common understanding across government departments of the threat posed by small arms: they are deemed to 'fuel conflict, crime, terrorism, human rights abuses and pose a major obstacle to sustainable development', with the demand for weapons often 'symptomatic of the underdevelopment of society' (Amos et al. 2003, 3; FCO no date; DfID et al. 2002, 3). In this understanding, poverty is linked to the emergence of conflict, which in turn further hampers development. Small arms, while not the cause of conflict, are understood to exacerbate it and fuel drug smuggling, criminality and terrorism, and to be linked to resource exploitation. According to the government, it is important to check the spread of small arms not only because of the suffering it causes in the developing world but because to do so is in Western interests as well: 'We are all linked up in the turbulence and disorder and criminality and suffering that results from this flow of small arms and light weapons across the world' (Short 2003). Small arms are thus understood to pose a threat not only to those living in the South but also to those in the North via the spillover of conflict and violence.

NGOs as nodes in strategic complexes of global liberal governance

Small arms have been a major growth area for NGOs since the 1990s and are the arms trade issue on which they have cultivated the closest relationship with governments, international organisations and IGOs. NGOs have more space to discuss small arms with government than other security problems because the smaller the weapon, the easier it is to 'sell' as an issue, according to a DfID official working on small arms. Another DfID official described how small arms 'can be a humanitarian issue, a women's rights issue, as well as a security issue, and these are areas where the government is more open'. From the NGO side, the lesson learned by the small arms community from the landmines campaign was to take issues out of the security field, and turn them into people-centred or humanitarian issues in order to broaden the debate beyond technical discussions between ministries of defence and trade. It has been important for NGOs to retain a humanitarian focus, according to an Amnesty campaigner; otherwise, they fear that the small arms issue will end up in a 'technical cul de sac', and then 'it doesn't get anywhere'. Small arms are more difficult to generate international controls on than landmines however, because of their deeply entrenched legitimacy as military, police and, in some parts of the world, civilian weapons (Clegg 1999; Garcia 2006, 5, 37; Hill 2006, 2).

The NGOs' approach is most closely associated with development agencies such as DfID, but it has become broadly shared across government. The UK government's main activity on small arms is channelled through cross-departmental Conflict Prevention Pools. The Global and Africa pools were established in 2001 to facilitate joint policies and coordinate work across government departments (Bell 2003, 3-5), as well as to 'influence broader government policy and mobilise support from other international actors' (FCO 2009). As part of the government's efforts at coherent, cross-departmental working, small arms control programmes are closely associated with the promotion of good governance and poverty reduction initiatives. The Global Pool took the lead on policy development, working 'closely with NGOs, governments, regional organisations and the UN to develop and implement targeted strategies for reducing the damage caused by armed violence and gun misuse' (Bell 2003, 33). This is part of a 'comprehensive and sustained response from the international community' across issues such as human rights, humanitarian aid, post-conflict reconstruction, development programmes, security sector reform, gun control and law enforcement (DfID *et al.* 2002, 4). In April 2008 the Global and Africa pools were merged into a single Conflict Prevention Pool, with a mix of regional and thematic streams, including security and small arms control.

The collaborative relationship with NGOs signals one of the notable features of action on small arms, namely the emergence of a network of actors that cuts across the supposed divide between the state and civil society. Such networks are characteristic of what Duffield (2001) calls the strategic complexes of global liberal governance that have emerged in response to the merging of development and security. State responsibilities are increasingly being privatised and subcontracted, with NGOs playing a key role. NGOs have formed networks and representative platforms amongst themselves, through which they gain access to decision-making fora, and donor–NGO funding relations have become increasingly formalised since the 1990s (Duffield 2001, chapter 3). The aid market is a two-way process, however: 'While many NGOs are dependent on government and IGO funding, their ability to monopolise local access and control information has given NGOs a strong role in policy formation' (Duffield 2001, 57). More widely, the emergence of shared policy assumptions has been crucial to the operation of these complexes (Duffield 2001, 73).

Action on small arms is illustrative of the emergence of such complexes. Amnesty International, International Alert, Oxfam and Saferworld, alongside research institutes such as the Small Arms Survey, all engage in research, advocacy and programming on small arms issues as part of a strategy of providing expertise to the state on policy making and project implementation. For example, International Alert and Saferworld provide training on conflict prevention to a range of actors across the UK government. They are both members of the Biting the Bullet team, which 'served as a semi-official advisor to the Foreign and Commonwealth Office' (Karp 2002, 181), with some of its research feeding directly into the UK presentation at the 2006 Review Conference, in particular on the impact of small arms on development, governance and security (Biting the Bullet 2006, 226-32). Since the conference, NGOs and academics have worked to promote implementation of the Programme of Action and create 'opportunities to discuss critical issues that proved controversial' at the conference (International Alert 2006c). A notable development in the role of NGOs as advocates on small arms issues came in 2006 when the director of Saferworld was invited on to the UK delegation to the UN Conference. This signals the ultimate insider role of Saferworld based on its expertise and respectability, and the willingness of the UK state to act in tandem with NGOs on small arms issues.

This UK-based network of actors also has a web of relationships with the South, primarily through NGO field programmes, funded by DfID and/or the Conflict Prevention Pool, predominantly in Sub-Saharan Africa, Eastern Europe and Asia. For example, Saferworld has long been active in East Africa and the Great Lakes region, helping develop National Focal Points and National Action Plans. These are designed to monitor the illicit trade, improve stockpile management and security, and improve national legislation on the domestic manufacture, production, export, import and transfer of small arms (UK Mission to the United Nations no date; DfID no date). Saferworld helps the National Focal Points to develop 'their roles, responsibilities, programmes of work and day-to-day operating procedures' (Saferworld no date, e). This work includes the development of National Action Plans, comprehensive programmes to tackle the small arms problem arrived at after a process of 'small arms mapping', research into the nature of the problem in each country and the means to address it (Saferworld no date, f). This work is undertaken with Saferworld's partner organisation, SaferAfrica, and in partnership with government and civil society representatives in each country.

Oxfam is also involved in small arms and conflict prevention work in East Africa, often alongside Saferworld, through its conflict and peacebuilding programme in Kenya and involvement in a UNDP project on small arms and development. Oxfam's work in Kenya is a country programme of Oxfam GB and revolves around peacebuilding and conflict management in northern Kenya, working with pastoralists, the government, NGOs and 'communities vulnerable to insecurity' (Oxfam 2006; also Waqo 2003). This work is based on Oxfam's observation that cattle rustling is made lethal by the influx of small arms; the organisation aims to develop a culture of peace, reduce demand for arms, and promote police training and firearms collections (Mungai 2006). Oxfam also funds national initiatives through the National Steering Committee for Peacebuilding and Conflict Management, within which the National Focal Point sits. It was established in 2001 and, with the help of Oxfam GB, had its secretariat functioning by early 2003. One of its key activities is the 'reduction of illicit small arms and light weapons in the country' (Oxfam 2003). According to one Oxfam campaigner, such community safety work is distinct both from Arms Trade Treaty campaigning and from humanitarian delivery work in countries like Sudan, where the atmosphere is inappropriate for raising small arms or armed violence campaign issues.

Small arms as an ideological battleground in global civil society

NGOs play an important role in building capacity for action on small arms issues in the South, funding and training Southern civil society organisations. They thus contribute to what is commonly understood in the literature as the globalisation of civil society activity, the border-crossing nature of civil society activism, or the rise of multilayered governance, as discussed in Chapter One. Such efforts are not restricted to liberal or cosmopolitan actors, however. Politically conservative, or even radically right-wing forces are also engaged in the terrain of struggle over the social meaning of small arms. The National Rifle Association, in particular, has been heavily involved in shaping the parameters of the debate on small arms in the US, and has made moves to internationalise its position, hampering progress on the restriction of civilian possession. The arms industry and its associated lobby groups, however, are in favour of an Arms Trade Treaty and tighter controls on the illicit trade as these would harmonise regulation and improve business prospects (e.g. BAE Systems 2008b; Defence Manufacturers' Association 2006). Small arms activism is thus an indicator of the ideologically charged nature of global civil society.

For Saferworld, the challenge is to find partner organisations with a broad constituency, links to the local community and the capacity to have people on the ground and coordinate activities. Considerable effort is put into maintaining a sense of mutual benefit between partners, although the existence of power relationships is acknowledged by staffers. Sustainability and ensuring partners have a degree of ownership in programmes pose a key challenge. Staffers working on regional programmes acknowledge that it can be difficult to ascertain whether organisations are genuinely committed; Saferworld is a Northern NGO with money, so Southern organisations see working with it as an opportunity to increase their profile, capacity and access to funding. As one regional staffer put it, while 'we're not there to build civil society organisations', capacity building is necessary so that Southern organisations can fulfil the functions required of them. Staff have found that the process of building capacity is often slower than the project allows, due to the need to improve financial management skills, individual staff members' capacities, and organisational management systems. Similarly, International Alert staff are aware of the risks involved, namely 'that you end up inviting the same old suspects – the articulate groups, those fluent in English'. According to one staff member, 'voices from the field are vital and in demand – their word gets taken as gospel – so it's important not to always get a member of the elite from the capital'. But when it comes to choosing partners to participate in an important conference, staff do not want to take a gamble, so they may rely on tried and tested partners.

NGOs recognise that civil society in the South has its own interests; activists may be trying to generate income for their own organisation or manoeuvre towards a job in government, for example. International Alert tries to get a balance of partners and, in countries in which all activity is seen as politically affiliated, it tries to work with a range of actors. As one staffer put it, 'You're never going to be seen as neutral, but at least you can try to get balance and be fair.' So International Alert has a policy of engaging with whomever it sees as key constituencies, even if they are known to use violence. This is not a hard and fast policy, as the safety of staff and interlocutors has to be protected, but actor mappings can reveal who the 'spoilers' are – and in some instances this can include the international community. Including the 'spoilers' is seen as vital: as one staffer described it, if you fail to engage and understand people's motivations for what they're doing, be it torture or extortion, 'then you're cutting out a whole side of the debate'.

IANSA engages in capacity-building work in a different way to International Alert and Saferworld. Rather than provide core funding to groups in the South and train them directly, it provides media releases and strategies, and information about regions and countries; it can also supply small amounts of funding for weeks of action and particular events. The London-based secretariat sees the role of the network as being to connect groups, and to generate a sense of solidarity that means they can then overlap with other campaigns. It also sees itself playing a more strategic role in facilitating Southern state participation in UN processes through its support for civil society. As a secretariat staff member explained, normally only European states have the resources to respond to UN Secretary General messages and consultations, but on small arms issues, IANSA members have taken the call to Southern governments and facilitated their participation in UN processes. So the network operates as a transmission belt between the UN and national governments in this view. There is a different relationship to government in the South compared to the North, according to the secretariat: Southern civil society actors are often more critical because their governments are more corrupt, so they see their role as criticising the government rather than talking to it. As seen from the secretariat then, the role of the IANSA network is to help build up the relationship between civil society and the state. According to this understanding, the role of civil society is to support Southern states and facilitate their participation in UN consultations and so on. Participation in UN processes is emblematic of these states' juridical sovereignty, but they are often unable substantively to exercise this right. And ultimately, the network's role is to promote a consultative relationship between NGOs and the state, to move from an attitude of confrontation to one of collaboration.

There are striking regional disparities within the IANSA network as well as different foci of member activism. The regions with the most members are Western Europe, North America and West Africa, which are significantly better represented than the Middle East and North Africa, Central Asia and Southeast Asia (see IANSA no date, d). This disparity has largely existed since the formation of IANSA: North America, Western Europe, Sub-Saharan Africa and Central and South America were the best represented regions in the late 1990s (Krause 2001, 22). The small arms network started off as a group of Northern (in particular British and American), security-oriented NGOs and academics coming out of the Cold War focused on arms export controls, according to one IANSA secretariat staffer (see also Karp 2002, 180). According to Krause, Latin America's strong civil society tradition and Sub-Saharan Africa's enmeshment with the international development community help account for these Southern regions' relative visibility in the network (Krause 2001, 21-2). In particular, Brazilian and South African NGOs (such as Viva Rio and Gun Free South Africa, respectively) have been key Southern influences on the small arms debate (Garcia 2006, 25). IANSA has attempted to be Southern-focused in its membership and there has been a deliberate relative lack of explicit leadership (Grillot et al. 2006). It has had 'global reach' since its inception, yet coverage has always been 'patchy', with East and northeast Asia, the former Soviet Union and the Middle East and North Africa underrepresented from the start (Krause 2001, 21). The most striking growth has been in Sub-Saharan Africa; 27 NGOs were represented in the early days of IANSA, while at the latest count the network has 134 members in the region.

According to a member of the secretariat, West European members tend to be high-expertise organisations focused on the technical aspects of small arms control, while few grassroots, anti-war, or peace organisations are involved. In the UK, for example, IANSA is the main link to the Gun Control Network. But UK laws are relatively tight and gun violence is a relatively marginal issue nationally. As one secretariat staffer characterised the situation, the UK is relatively safe and so the issues are more pressing elsewhere. In North America, civil society tends to be focused on the domestic debate, in contrast to West European groups, which are mostly focused on the South. Washington-based groups have a strong history of involvement in gun and small arms control debates, but the Bush administration meant a decline in interest and funding for work on reducing gun violence. While there are thus significant restraints on US involvement on gun/small arms work – US groups would never use the IANSA logo of a gun with a red line through it, for example, according to the secretariat – Amnesty USA is very involved on the Arms Trade Treaty. The strong presence of West European and US groups within IANSA has been criticised from within the network: a survey of IANSA participants found that there is a strong view within the network that it is dominated by a few Northern organisations – Amnesty International, Human Rights Watch, International Alert, Oxfam, Saferworld, and Small Arms Survey. These groups do have greater resources, skills and information, but also 'are often charged with drowning out the voices of other actors' (Grillot et al. 2006, 74).

In the South, the regions that are very active, such as West Africa, feature significant numbers of grassroots campaigners that have been working on small arms issues since before the advent of the formal IANSA network. IANSA plays a facilitating and, when necessary, representative role for Southern civil society. In order to get the concerns of the South reflected in the North, IANSA brings members to UN and EU meetings to allow their voices to be heard directly. If Southerners cannot participate, for resource or other reasons, Northern activists represent them, as one member of the secretariat put it. However, such a view is contested from within the network: survey data suggest that, while some participants feel that Northern groups should 'provide information, training and exchange opportunities', they also 'must allow space for alternative views' and 'must not manipulate the positions of NGOs in the South to serve their needs' (quoted in Grillot et al. 2006, 75). The formation of IANSA has led to a rise in the number of civil society organisations around the world working on arms issues, and has galvanised members into working on the Arms Trade Treaty. Its membership has grown quite substantially since 2003, although the secretariat understands this in terms of existing human rights or development organisations introducing Control Arms activism into their existing work, rather than the establishment of new groups simply to be part of IANSA.

The South-driven nature of the network determines IANSA's focus: while some Northern members feel IANSA is doing too much Arms Trade Treaty work at the expense of issues such as NATO enlargement, the secretariat says the Arms Trade Treaty is the priority identified by its Southern members, or at least by a majority of the network. The reason given for this is that in the South, the problem of the arms trade *is* the small arms trade: most Southern countries do not have nuclear weapons, and do not have much in the way of conventional weapons, so the disarmament people *are* the small arms people, as one secretariat staffer described it. And the role of guns in crime is a big issue in the South: in regions such as Latin America, guns or small arms are much more of a crime issue than a military issue; and when Africans speak at international meetings, secretariat staffers describe how they often repeat the refrain that small arms are 'the real weapons of mass destruction'. This is echoed by other NGOs, scholars and practitioners, who emphasise the particularly destructive role of small arms: put simply, 'guns are the weapons that kill the most people' and they are 'disproportionately harmful' (Peters, in IRIN 2006, 36). An estimated 90 per cent of fatalities in conflicts since 1990 have been among non-combatants (Clegg 1999); and small arms 'account for the overwhelming majority of deaths in conflicts since 1945' (Krause 2001, 8).

Some regions of the South are under-represented within IANSA. According to the secretariat, the main problem in the Middle East is IANSA's lack of an Arabic speaker: while there is interest in the Middle East and the members there are very good, there are not many of them and the secretariat has not been able to facilitate membership growth there. In Southeast Asia, the gun issue is very tightly linked to trafficking and terrorism and is perceived as a national security issue. And in China, the human rights organisations that exist in other parts of the South 'just aren't there', according to the secretariat.

Thus, global civil society activism on small arms is unevenly distributed geographically, suffers from competition within the IANSA network (Grillot *et al.* 2006), and is also contested ideologically from outside. As already mentioned, the National Rifle Association and conservative gun rights activists in the US have also been heavily involved on the small arms issue. Like the pro-control network associated with IANSA, pro-gun-rights activists have lobbied national governments, promoted grassroots campaigns, formed links with overseas organisations, and created transnational organisations; in their case, the World Forum on the Future of Sport Shooting Activities (WFSA) (Bob 2007, 199-200). The WFSA only has 38 members, compared to IANSA's more than 700, and gun control activism tends to be more South-focused, although the NRA was involved in international outreach in, for example, Brazil, lobbying prior to the referendum that defeated a ban on civilian gun ownership (Bob 2007, 200). The NRA exercised considerable influence over the US government at the 2001 UN small arms conference: the US delegation to the UN Conference included three NGO members, all from the NRA (Erickson 2007, 8). The goals of the Programme of Action, limited as they were compared to what the gun controllers were calling for, were 'the result of US government "redlines" supported by the NRA' (Bob 2007, 200). The NRA's presence was in marked contrast to 'the amazing invisibility of America's domestic gun control activists' such as Hand Gun Control and the Coalition to Stop Gun Violence, whose absence 'simply gave the NRA a monopoly at the conference' (Karp 2002, 188, 191).

International action on small arms has been heavily coloured by the US domestic debate about gun control, pitting a conservative vision of individual rights, protection against the state, and an emphasis on national state sovereignty, against a liberal internationalist vision of the rule of law, state monopoly on violence, and international cooperation. The NRA argues that 'Free people elect good government' and that freedom involves being allowed to 'embrace American Constitutional freedom in the Bill of Rights', complete with the right to bear arms. Individuals should retain 'their right to protect themselves from the predators, the evil doers, the killers and the genocidal governments' (Peters and LaPierre 2004). More widely, gun rights proponents argue that citizens should retain the right to bear arms as protection against the state (Kates 2003); that it is states that kill people much more than criminals or insurgents (Kates 2003); and that 'violence has been a central aspect of some of the most significant and productive events in US history' (Perlstein 2003, 314). For them, non-state actors 'may be oppressed groups attempting to secure their legitimate rights in the face of a tyrannical government and should therefore be entitled to receive arms' (Kopel 2003, 319).

Gun controllers reply that the aim is not a ban on civilian ownership but 'to disarm guerrilla armies and ameliorate the catastrophic waves of gun crime that often follow a war' (Karp 2003, 310; also Small Arms Survey 2009, chapter 5); and that small arms control proposals would curb the supply of weapons to state as well as non-state actors that violate human rights and international humanitarian law (Bondi 2003, 322). IANSA itself says that small arms control 'isn't about a gun ban. We're talking about some moderate measures to reduce the illicit traffic in guns...We're talking about bringing the arms trade under control, to stop guns getting into the hands of criminals and of drug gangs and of human rights abusers' (Peters and LaPierre 2004). However, the IANSA founding document has as a goal the reduction of 'the availability of weapons to civilians in all societies' (IANSA no date, b). This is one element of a wider vision around the role of the state and its relationship towards its citizens and residents. For IANSA, 'the way to get freedom, the way to have democracy, is to have stronger institutions and the rule of law'. The social contract in which individuals form societies and governments – and, by extension, surrender the use of violence to the state – 'operate[s] better than a whole lot of individuals making up their own rules, taking the law into their own hands' (Peters and LaPierre 2004).

Assessing NGO activity

NGOs have devoted significant efforts to and exercised considerable influence on the issue of small arms and conflict. They have criticised states for controversial practices, set out rules to govern arms transfers, and been involved in Southern efforts to reduce weapons proliferation and gun violence. NGOs have been relatively successful in moving small arms out of a state security framing and into a development, humanitarian and human rights framing, facilitating international action. While the shift to a human security agenda has been crucial to the successes of small arms control such as they are, and some headway has been made on issues such as regulating illicit brokers, destroying surplus weapons and improving marking, tracing and record-keeping (Garcia 2006), 'harder', state-sanctioned definitions of security have survived relatively unscathed.

NGOs and development agencies exercise considerable influence on issues of conflict, security and development, but only as long as these do not threaten the interests of arms capital or those elements of the state with which it is most integrated, or historically entrenched understandings of state sovereignty and national defence. For example, in the case of Tanzania, DfID and NGOs are all taking the lead to address the country's small arms problem, and Tanzania has been a pioneer on small arms, with its government becoming the first in the Horn of Africa to develop a National Plan (Eavis 2002, 258). Yet DfID and the NGOs were powerless to prevent the Tanzania air traffic control system deal discussed in Chapter Five, for example. If anything, action on small arms creates the impression of benevolence, allowing the wider arms trade and state-centric, military definitions of security to continue relatively unchallenged.

In addition, the human security framing itself needs to be subject to critical scrutiny. The argument that the spread of small arms is a contributory factor to conflict, insecurity and underdevelopment, and that holistic solutions are needed in response, illustrates the ways in which conflict has been incorporated into mainstream development policy. According to this understanding, development cannot take place without security, and security cannot be achieved without development (Duffield 2001). This approach is understood by its proponents as progressive, as a shift away from state-based security and towards human security. In response, a broad good-governance approach has emerged: whilst dangerous, small arms availability is a 'predisposing' rather than a 'fundamental' cause of underdevelopment (Small Arms Survey 2003, 128), and underdevelopment is deemed 'open to remedy and demands engagement' (Duffield 2001, 114). Small arms projects are thus one measure in a broader programme of social transformation.

However, while the illicit trade in small arms is associated in policy and much academic discourse with crime and underdevelopment, it is not clear that conflict does destroy development. As Duffield argues, 'the transborder networks that support organised violence in one location have encouraged autonomous and resistant processes of *actually existing development* in other areas…[in the]…spaces and lacuna created by structural adjustment and globalisation' (Duffield 2002b, 160, emphasis in original). This means that, rather than signalling an absence of development, such seemingly criminal or illegitimate activities are a form of development, albeit not one recognised as progressive by aid donors. Such development is intimately connected to processes of state formation, the legacy of colonialism, and the global capitalist economy. That is, 'Violence and war should not…be seen as oddities, distortions or distractions but should be regarded as closely connected to progress and development' (Cramer 2007, 45). Economic globalisation as promoted by state and capitalist elites has been significant in creating the conditions to which alternative development is a response. These 'decentralised shadow economies' foment 'destabilising forms of global circulation' (Duffield 2005, 143, 156; also Duffield 2001, 5) but, rather than signifying social regression, the violence associated with this is better understood as 'reflexive modernisation'. Although they are non-liberal, the types of social relation being formed are those of 'autonomy, protection and social regulation' (Beck, quoted in Duffield 2002a, 1055; also Sörensen 2002, 13, 16).

Such processes do not signal an absence of development or governance; rather, they are forms of 'actually existing' processes, even though they are understood as regressive by donors. The distinction between 'cheap war' based on violence with small arms and 'expensive wars in which civilians are maimed or destroyed with sophisticated laser-guided weapons' has political but not analytical value (Richards 1996, xx). Much of the development literature views conflict as 'temporary and a universally bad thing for everyone involved' (Jackson 2003, 148). Organised violence has devastating effects on its victims, but 'for those groups in whose name it is carried out, actors are saviours and protectors rather that criminals or manipulative elites' (Duffield 2002a, 1060). Thus, whilst violent conflict is undoubtedly occurring, it can only be understood with reference to the context within which it occurs. The use of force in such contexts may well have a very different meaning to that imputed to it by donors, but this should not in and of itself mean that it is illegitimate or senseless. Whilst not recognised as 'good' governance, violence plays a role in forms of governance nonetheless. Violent, illiberal social relations are a *form* of governance, rather than the absence of it.

Many parts of the South feature non-liberal models of political authority and because these often involve violence, they attract the attention of aid donors and NGOs, who tend to understand it through liberal lenses. Their conception of violence as a pathology leads them to favour solutions involving social transformation: whole packages of measures ranging from the psycho-social to the community level and national good governance strategies. But understanding the emergence of the illicit trade in small arms, and the wider 'new wars' and 'failed states' to which it is endemic, as 'a consequence of long histories of colonial and postcolonial interaction with the West' (Barkawi and Laffey 2006, 347) means that we need to think about processes of interaction and relationships, as opposed to seeing these problems as internal to the South.

Small arms emerged as a multilateral security problem as a result of the 'changed matrix of conflicts in the post-Cold War world' and 'expansion of multilateral peace and security operations' in such conflict areas (Krause 2001, 10). However, 'Very little, if anything, objectively changes in the early 1990s', suggesting that the increase in concern about small arms was due predominantly to the creation of expert knowledge and changed normative space rather than to a change in the nature of weapons flows, their use in conflict, or their impact on major actors' security interests (Krause 2001, 17). Thus, rather than simply a description of the world or an objective policy response, international action on small arms and the conflict prevention agenda associated with it can be understood as a 'bugle call for collective mobilisation', playing 'a symbolic rather than an informational role' (Duffield 2001, 116) and legitimising a 'will to govern' that is linked to the 'long-established reforming urge within liberal societies' (Duffield 2002a, 1052-3). It stems from a liberal impulse to do good in the world, opening the way for complete social transformation in various parts of the South.

The international small arms agenda is a key means of European engagement with the non-European world. It serves as a mechanism of mutual constitution, in that attempts at social transformation and state-building in the South go hand in hand with the construction of a Northern self as benevolent, concerned about the impact of globalisation on the poor, having a desire to prevent conflict, and so on. Small arms control is thus indicative of Eurocentric security studies that 'regard the weak and the powerless as marginal or derivative elements of world politics, as at the best the site of liberal good intentions or at worst a potential source of threats' (Barkawi and Laffey 2006, 332).

The emphasis on the illicit trade in small arms is a key way in which the North and South are differentiated from each other: 'we' have control over state borders and weapons stockpiles in our modern, territorially bound states, while 'they' have criminal and insurgent networks, anarchy, corruption and a general absence of good governance. Two sets of practices are facilitated by this, both based on the conflict–security–development argument: the shoring up of state monopolies on violence, and removal of weapons from societies. The former is facilitated by attempts to promote state controls over arms exports, and by their work with Southern states in improving stockpile management, updating legislation, destroying surplus weapons and so on. The latter is facilitated by field programmes in the South, with their focus on community safety and DDR. These respective supply- and demand-side approaches are complementary: for example, Amnesty's criticism of the excessive use of force by the Ugandan government in its programme to disarm pastoralist communities (Amnesty International 2008) is the supply-side counterpart to the work of Oxfam and Saferworld on community safety in the region. NGOs are promoting processes of the removal of weapons from societies and the creation of a state monopoly on violence, but they want this to be done without excessive force. As the Small Arms Survey argues more generally, 'The only promising solution appears to be state-building and restoration of the rule of law' (2009, 167). However, in the case of pastoralists in and around Uganda and Kenya, recent attempts at disarmament as a means to development have not only failed in their own terms, they also replicate older practices of state building from the days of the dissolution of empire (Knighton 2003).

In addition, the geographical areas on the receiving end of small arms interventions are heavily circumscribed. In practice, the spatial metaphor of the borderlands is selectively operationalised, with small arms activism occurring mostly in regions such as Sub-Saharan Africa and the Balkans, while the Middle East, and South and East Asia receive less attention. NGOs put this down to the issue being too firmly ensconced in a national security framing, and to the weakness or absence of civil society in these regions. Rather than signalling a lack or a weakness, however, it is more fruitful to think of the small arms control agenda as a counterpart to the wider arms export agenda. As with the human rights and development agenda discussed in Chapter Five, the conflict prevention agenda and ongoing arms exports are two sides of the same coin. There is an attempt both to contain conflict in certain parts of the South and to contribute to the maintenance of state coercive capacities. Where possible, weapons are removed from societies and liberal social relations are promoted; elsewhere, where it would be too risky to attempt such measures, more repressive incorporation is preferred, through the export of weaponry to states. Small arms clean-up programmes thus tend not to be carried out in regions to which major arms-exporting states have a record of exporting weaponry, such as the Middle East. And as moves to reduce military spending and procurement apply to (parts of) the South but not the North, so too the community safety approach is a South-oriented agenda, with a focus on reducing levels of civilian possession. The removal of weapons from Southern societies is politically permissible in a way it is not in the US, in particular.

It is important to note that the conflict–security–development agenda is not simply being imposed on the South by the North. There is significant demand from within the South, by both governments and NGOs, for tougher controls on small arms. And although 'a handful of governments in the South had sought for several years to focus the United Nations on the dangers of the illicit small arms traffic', the issue did not capture the international security imagination until 1995, when UN Secretary General Boutros-Ghali called for 'micro-disarmament', giving the issue 'traction'

(Lumpe *et al.* 2000, 7). And from the earliest days of international action, NGOs have understood that any small arms campaign 'must also account for the relationship between North and South'. That is, 'a truly effective light weapons campaign cannot simply be seen as a way of "disarming the Third World"' (Clegg 1999, 50). However, this should not blind us to the liberal assumptions underlying the conflict–security–development nexus, nor to the ways in which international action has developed in practice, creating the South as site of intervention. More generally, there are also wider processes of military build-up in the South, involving the purchases of major conventional weaponry, which are encouraged by a Northern-dominated global military culture yet not often the focus of NGO activism.

Conclusion

At one time or another, all seven of the NGOs discussed here have worked on small arms. They share an understanding of the risks of conflict, and the threats to development and security posed by the spread of this category of weaponry. There has been a shift over time, however, with BASIC and CAAT moving away from small arms to focus on nuclear weapons and the state–capital relationship, respectively. The other NGOs have increased their work on small arms, establishing field programmes in the South and building their experiences of this into their international advocacy via the United Nations. They articulate both supply-side and demand-side arguments about the spread of small arms and are pursuing an international Arms Trade Treaty, which has a special emphasis on small arms, despite being aimed at the conventional trade as a whole.

NGO activity on small arms issues has two main, interrelated effects. First, it serves to remove weapons from non-state actors in the South. Efforts to remove weapons from non-state actors in the North are not attempted, partly because of the US state's absolute refusal to consider increased restrictions on civilian possession beyond its current regulations, and through a narrative of the link between small arms and conflict that only pertains to the South. Second, the NGOs' work is an attempt to shore up, or create, Weberian ideal-type state sovereignty. NGO activity in helping Southern states develop legislation, control their borders and manage weapons stockpiles signals an attempt to instantiate modern, national territorial states that have a monopoly on violence. As is discussed further in Chapter Seven, not only is this an idealised image of the state within Europe, but this activity also occurs in postcolonial contexts where the modern national territorial state has never been in operation in the same way.

Small arms programmes demonstrate the emergence of a networked form of governance in which NGOs act alongside states and IGOs. While states ultimately control the overall agenda, NGOs have become key intellectual and project partners, with Saferworld, International Alert and Oxfam, in particular, significantly integrated into DfID's work. In a sense, this represents the privatisation and de-territorialisation of state influence (Duffield 2001, 72). It also represents the uneven internationalisation of public policy in the South: NGOs such as International Alert and Saferworld are heavily involved in shaping state policy and promoting implementation in countries across East and West Africa, as well as Eastern Europe.

NGOs are conscious of this transformation, although they would not describe it in the same terms. Their efforts at advocacy and capacity building are based on the argument that: 'Assistance is slowly beginning to move away from typical models of Northern patrons assisting developing and transitional states – towards a stronger web of cooperative assistance relationships at all levels' (Biting the Bullet 2006, 235-6). NGOs are therefore actively participating in the transformation of governance relations, celebrating what they understand as the end of Western imposition and a process of partnership between governments and NGOs in the North and South. While their activity may not signal imposition – not least because the conflict–security–development nexus is heavily emphasised by some Southern governments and NGOs – there is nonetheless a process of the diffusion and promotion of liberal values and practices. One of the implications of this merging of relationships between NGOs and governments is that the narrow question of whether NGOs influence states or vice versa becomes less of an issue. It is more productive to analyse the co-production of small arms as a problem and the cooperative measures to tackle them. Importantly, this interaction takes place on small arms issues but not on the wider trade in conventional arms.

Small arms measures promoted by the UK government are part of wider processes that incorporate SSR, DDR (disarmament, demobilisation and reintegration), good governance and other such initiatives. While such projects are imbued with the language of participation, holism and an awareness of the historical and political reasons for the spread of small arms, taken together they signal an attempt to remodel the South in an idealised, ahistorical and ultimately depoliticised image of the North. Small arms clean-up programmes allow leading arms-exporting states to appear benevolent and legitimise the wider trade in arms. Small arms control efforts and the ongoing export of weaponry (in the form of small arms and other conventional weaponry) are thus two sides of the same coin: the maintenance of coercive capabilities, especially where regime stability must be maintained; and conflict prevention as a pacifying measure. NGOs are key actors in these processes, reproducing dominant discourses and acting as sub-contractors to the state. Such activity has a number of unanticipated effects: naturalising and operationalising dominant liberal understandings of conflict that fail to take political grievance seriously; facilitating intervention in the South in the form of small arms clean-up programmes and associated attempts at good governance; and facilitating the construction of a benevolent and charitable Northern identity, part of the mutual constitution of the North and South through small arms issues.

7 • NGOs, Global Civil Society and the World Military Order

The NGO community has played a significant role in putting arms trade issues on the international security agenda since the end of the Cold War. Individually and collectively, through a variety of strategies, NGOs have raised public and elite awareness about the effects of the arms trade, and contributed to elements of policy change, although the issues on which they have had influence are heavily circumscribed. More broadly, their practices are interventions in international relations: NGOs contribute to the production of the arms trade as a problem and they facilitate particular types of responses to it. Previous chapters have demonstrated how this plays out in relation to the main trends of the arms trade and the contested and often problematic effects of NGO interventions. As a result, NGO activity on the arms trade raises significant questions for our understanding of civil society, on all four counts identified in Chapter One: as separate from both the state and market; as progressive; as global; and as non-violent. The preceding analysis raises questions about the adequacy of this dominant understanding.

Rather than understanding civil society as substantively separate from the state and market, we can more fruitfully think in terms of dual networks of actors, one comprising NGOs and the development agencies, the other comprising arms capital and the defence-industrial branches of the state. Countering mainstream claims that NGOs are bearers of progressive values, they can be understood as propagators of hegemonic liberal ideas around development, human rights and conflict, and the practices associated with these ideas are better understood as imperial in nature. The NGOs' various strategies are conventionally understood to be complementary, although the reverse is rarely considered. However, the question remains of whether the social forces exist for a more transgressive vision to take hold. While civil society is held to be becoming increasingly global, its development demonstrates significant ongoing North–South hierarchies. And, rather

163

than understanding global civil society as proponents of a pacific world order, NGOs' role in trying to reduce certain forms of violence while sidelining others means they are better understood as unwittingly facilitating a profoundly hierarchical and violent world military order.

Dual networks

The integration of arms capital and certain NGOs into state structures challenges dominant conceptualisations of global civil society as substantively separate from the state and market. The close relationship between the more insider NGOs and DfID is generally understood to be an indicator of their effectiveness as it routinises NGOs' relationships with the state, allowing them to make suggestions that are taken seriously and to shift policy, albeit in incremental ways. The UK Working Group/Control Arms NGOs have been highly effective in this sense. They were pivotal in the emergence of the EU Code of Conduct and its subsequent adoption as a legally binding EU Common Position; the UK state has incorporated NGO terminology and wording into regulations on torture equipment, brokering and transportation issues; some NGO suggestions for improvements in annual reporting on arms exports have been adopted; and NGOs are working alongside the UK government in pursuit of stricter international regulation of transfers through the Arms Trade Treaty. NGO suggestions have become normalised within government over the years, such that their eventual adoption has become common sense. NGO integration is particularly pronounced on small arms, with International Alert, Oxfam and Saferworld acting in partnership with DfID to operationalise small arms clean-up programmes in the South and to promote international action to restrain the illicit trade via the UN Programme of Action. On small arms issues, NGOs have also been involved in the transformation of state power in the South, playing a key advisory role in legislation reviews, mappings, policy design and implementation, capacity building and so on.

Despite these successes, however, NGOs have shifted neither the overall orientation of UK policy nor the wider arms trade. The scale of the international arms trade, export patterns and domestic procurement patterns have not changed significantly since the structural shift at the end of the Cold War. The UK government takes NGO advice when it does not threaten defence-industrial and 'hard' state security interests to do so, but ignores NGOs and other critics when it considers that necessary. For example, in 2004 the UK parliamentary committee responsible for scrutinising arms export policy was critical of the government's explanation of the change in assurances from the Indonesian government concerning the use of UK-supplied military equipment; the provision of information and policy suggestions to this committee is a key tactic for most of the NGOs. However, in response, the government simply stated that it 'does not accept' the committee's conclusion (Secretaries of State for Defence *et al.* 2004, 9). While NGOs have been pivotal to the committee's efforts to hold the government to account, their ability to discharge this responsibility fully or generate a change in policy is questionable.

The state's orientation towards the arms trade is in part shaped by its relationship with arms capital, which is itself embedded in deeply entrenched practices of state sovereignty and national defence. In the UK case, the integration of arms capital into the state, primarily via the revolving door and military advisory bodies associated with the MoD and BIS, means that the parameters of defence-industrial and arms export policy are set by an elite group of state and industry actors. This significantly weakens the relative autonomy of the state, lending weight to claims for higher military spending and support for arms companies and arms exports. In the South, while NGOs have had influence on small arms issues, they have had less impact regarding transfers of major conventional weaponry. Such transfers, with the associated relationships of patronage and often bribery that accompany them, serve to build the material and institutional basis of the state and embed arms capital into state structures. Internationally, NGO activism in favour of controls on small arms and the wider conventional trade finds favour as long as it is not seen to threaten state sovereignty or defence-industrial interests, which are often conflated when faced with a challenge from NGOs.

The dual networks facilitate two complementary sets of practices that favour arms capital and reformist NGOs, respectively. While NGOs have achieved increased access and influence under New Labour, shaping and reworking what it means for the UK to be a 'responsible' arms trader, state collaboration with NGOs on issues such as small arms, information provision, controls on brokering and so on plays a role in legitimating the UK's involvement in the wider arms trade. It allows both the government and industry to be perceived as benevolent actors in international relations, as they are both in favour of an Arms Trade Treaty and efforts to regulate the illicit trade in small arms. NGO successes on small arms must be contextualised in terms of the integration of arms capital into the state, which creates a permissive attitude for the wider production and export of military equipment. The post–Cold War and post–9/11 security environment of anti-terrorism and conflict, security and development is marked by both the promotion of high-technology, baroque (Kaldor 1983) military equipment and attempts to regulate the circulation of particular technologies such as small arms.

BASIC and CAAT stand outside these two networks to an extent, albeit in different ways. CAAT, through its isolation from the main arms control community, is not part of the DfID–NGO network. However, its outsider activity has made it a target for intervention by arms capital: it is unwillingly enmeshed in state–capital networks via its infiltration at the hands of private investigators paid by BAE Systems. BASIC is unusual in combining a highly insider, reformist strategy with a hard-hitting political critique that resonates with the vision put forward by CAAT and the wider peace movement. While it adopts a consensual strategy, it also opens avenues for wider critiques to be heard.

As a general rule, the structural disciplining of NGO activity through integration into the state and capitalist structures of civil society means that a consensual attitude is more than just a consciously chosen strategy: it is conditioned through NGOs' positioning within civil society, their access to funding and policymakers, and so on. As Hopgood (drawing on a different set of theoretical resources) argues in relation to Amnesty International, 'the IS [International Secretariat] and its toiling staff have not been unwitting ciphers or transmitters of iron laws'. We should give 'a qualified independence to agents while accepting that much of what they think, believe and value is the product of historical structures' (Hopgood 2006, 216). There may be a tension between the views of individuals working within NGOs and the arguments that their organisations can articulate. Organising as an NGO conditions collective actors to work in particular ways, one of which is generally to push for incremental reformist changes. This is what makes BASIC's attempts to create room for manoeuvre so interesting. As discussed below, however, it remains open to question whether the social forces exist for CAAT and BASIC's more transgressive vision to gain momentum.

The activities of these dual networks should be understood as complementary rather than antagonistic. That is, small arms control in the South, on the one hand, and the ongoing arms trade (including both intra-Northern military production and trade and exports to the South), on the other, are two side of the same coin. Small arms control is one element of the wider promotion of good governance, conflict prevention and poverty reduction strategies that come under the rubric of the post-Cold War shift to the promotion of polyarchy. Moves towards polyarchic social relations never entail the complete demilitarisation of society, however; the state always retains its coercive capacity (Robinson 1996, 65-6). And this coercive capacity is provided, in part, by arms transfers. This dual impulse helps explain US reluctance to engage in the Arms Trade Treaty and small arms control processes. The US state is happy to allow other states, particularly European ones and Canada, to take a lead on arms control, as long as it does not interfere with the US's coercive capacity and ability to transmit this internationally through support for foreign social forces.

NGO activity on small arms, the issue on which they can claim to be most effective, is emblematic of what Duffield (2001) refers to as the strategic complexes of global liberal governance. However, the privatisation and de-territorialisation of state influence that accompanies this is restricted to this issue and is itself geographically circumscribed. That is, small arms control and the foreign/development policy of which it is a part only takes place in certain parts of the South, and NGOs have not been able to get anything like the same levels of access and influence in relation to defence policy and exports of major conventional weaponry. Duffield acknowledges that these two sets of practices (which he refers to as geopolitics and biopolitics) are 'complementary, interdependent and work together to lesser or greater degrees' (Duffield 2005, 413). The analysis of NGO activity on the arms trade as a whole, rather than simply on small arms, suggests that the trans-formation of state power through the rise of NGOs and the conflict–security–development nexus is only part of the story of international relations in the 21st century. States, acting in tandem with arms capital, continue to promote authoritarian and illiberal practices in significant parts of the South, so the two networks and two sets of practices need to be understood in relation to each other.

Progressive values, liberalism and the normalisation of imperial practices

As key agents of global civil society, NGOs are widely understood to be the bearers of progressive values. At first glance, NGO efforts to promote humanitarianism, conflict prevention, peacebuilding, human rights and development seem to provide evidence of this. NGOs have had to battle against recalcitrant governments, adopting strategies of either persuasion or protest in an effort to get their concerns more or less firmly ensconced within institutional practices. This difficulty notwithstanding, the liberal underpinnings of reformist activism require scrutiny, as they may actually stand in the way of more transgressive action.

The reformist arguments that controversial transfers are an aberration in an otherwise benevolent foreign and development policy, and that it is in the common interest to raise normative standards to ensure the smoother running of international affairs, are key liberal tenets of international relations. Strategically, they make sense as a means of encouraging states to exercise tighter controls: they are persuasive rather than confrontational arguments. The assumption that the international architecture of foreign and development policy contain the promise of equality and progress, within which the eradication of poverty and enjoyment of full human rights by all is difficult but possible, rests on a belief in the pacific nature of capitalist development. It has no room to consider 'controversial' exports as the logical expression of foreign and development policies that rely on liberal social relations where possible but illiberal authoritarianism where necessary, or as an expression of capitalist development that necessarily includes the creation and spread of poverty, and abuse of human rights in the South in the process. And it has no room for the proposition that 'development' has historically been a violent process.

A key effect of the belief in the pacific nature of capitalist development and of liberal assumptions behind the role of organised violence in society is to reproduce the South as a site of intervention. Not only does it sideline the organised violence that is regularly visited on the South by the North, it also obscures the North–South relations that constitute the actors we tend to think of as unitary, sovereign states. These relations often revolve around support for elites through arms transfers as part of the process of state building, in processes that have not ended just because Southern states have won their formal independence. And while violent conflict is more prevalent within the South than within the North, the contradictions that produce it are often rooted in historical North–South dynamics as well as in intra-Southern causes. While a focus on Southern agency is important, the North–South structures of international relations play a significant role in producing particular forms of agency, both in the North and the South. NGOs' focus on the South is operationalised through a panoply of measures that are supposed to be universal but function to scrutinise Southern behaviour, as seen through the proposed methodologies for the Arms Trade Treaty. As Hopgood argues, 'The only practices that can plausibly be called international norms in the modern era are…those that accord with liberalism and the hegemony of the West' (Hopgood 2006, 216). The liberal claim to universalism is not a disinterested claim, but its greatest success has been in presenting itself as such.

Contextualising the arms work of the larger NGOs in relation to their overall orientation is further evidence of this. Amnesty International, for example, 'does not criticise systems of government as such, only the actual repression under such systems' (Amnesty International Mandate Committee, quoted in Hopgood 2006, 93). Its work is based on a liberal claim to detachment, objectivity and universalism. This is what gives it moral authority, but simultaneously means that the organisation 'lacks political relevance. When the issue is the systemic and systematic maltreatment of a class of people in the here and now, appeals to good conscience – in the face of dictatorship, violence against women, slavery and torture – lack conviction. To secure change you need to mobilise social power' (Hopgood 2006, 105-6). Amnesty is thus caught in something of a trap: its successes are based on an impartial, moral authority, which is precisely what stops it taking a side and making an argument that would *end* the violations that concern it. Oxfam, meanwhile, operates with a liberal internationalist belief in free trade, and a depoliticised model of development and globalisation. It understands these phenomena as neutral processes that could work for the betterment of humanity if only politicians would stop bringing politics in and skewing the processes in favour of the rich world (Berry and Gabay 2009). Both organisations attempt to address the intensely political question of the arms trade through apolitical stances and means – despite their willingness to criticise government policies and individual practices. What is missing is a wider critique that would allow them to play a role in ending poverty and human rights violations – but to make this critique would render them beyond the pale strategically.

Again, the partial exception to this is the informal alliance between the positions of BASIC and CAAT. Their focus on the North starts to get at the impetus to the arms trade, rather than simply its effects. Their proposed solutions start within the North, too. But to an extent this is a question of strategy within the NGO world: it is feasible that some Saferworld, Amnesty and Oxfam staffers would agree with (some of) the arguments being made by BASIC and CAAT, but their organisations do not make these arguments the focus of their strategy. The organisations believe that change can be effectuated by tapping into mainstream understandings and by working from current state positions. BASIC and CAAT, meanwhile, both agree with the Control Arms NGOs about the effects of the arms trade on the South. But they choose not to work on it, partly to avoid duplicating effort within the NGO world, partly because of their belief that the impetus to the problem lies elsewhere and, in BASIC's case, because the use of nuclear weapons is a greater threat than the use of conventional weaponry, including small arms. In taking such a position, however, CAAT and BASIC run the risk of appearing out of step with the most pressing issues, which are, in part, produced by the activism of the other NGOs.

NGO activity on the arms trade tends to reproduce liberal conceptions of the arms trade, development, human rights and conflict prevention. Although they are often critical of government practice, NGOs play a significant role in legitimating, buttressing and further entrenching liberal understandings that are inadequate for understanding organised violence in the contemporary world. Thus, rather than a progressive realm from which a more emancipatory future might spring, they are better understood as a transmission belt for imperial perspectives and practices, as part of a hegemonic constellation of social forces that perpetuates consensual domi- nation. There is a particular irony in this study of NGO activity given its empirical focus on the arms trade: Gramsci understood consensual domination to be backed up by coercion, and here we see NGOs playing a role in generating consent *for* some types of coercion and militarisation.

Insiders, outsiders, cumulative impact and social forces

The NGOs' various strategies, on a spectrum of insider and outsider activity, are widely seen as complementary, in that outsiders create the con- ditions that facilitate insiders making improvements. While this is a difficult claim to assess, there is evidence for it, such as in the closure of DESO in 2007. More widely, the activity of CAAT in calling for arms embargoes on states such as Indonesia and Israel, and its case of judicial review against the UK government in relation to allegations of bribery in arms sales to Saudi Arabia, allows the suggestions for change by the other NGOs to seem more reasonable and thus feasible. However, the substantive impact of these changes – facilitated by insider advocacy and backed up by outsider pressure – remains unclear, if and when they occur. The closure of DESO, for example, was a change in the machinery of government rather than policy, government-to-government deals remain under the control of the MoD, and so on. And despite claims to tighter licensing policy towards states that have records of human rights violations, UK-supplied equipment has still been deployed in campaigns of military repression in Indonesia and

Israel but also beyond. More widely, while weapons transfers to Middle Eastern states – a key recipient region for UK arms exports – are less often used directly in internal repression, their role in consolidating and reproducing the institutional and material basis of states continues.

There are thus grounds to ask whether the inverse might be true: whether insider strategies contribute to the conditions that undermine the effectiveness of outsiders, beyond whatever limitations the latter might have internally. Transformatory change requires incremental steps in order to come about, but the ideas behind incremental policy proposals must be transgressive if they are to contribute to counter-hegemony and the realisation of alternative social relations. Yet the policy solutions put forward by insider groups necessarily take the existing framework as given and thus naturalise it. This stems from their strategy of generating a consultative relationship with government and their liberal understanding of the problems associated with the arms trade.

As a self-identified transformist outsider group, CAAT does not want to generate consultative relationships with government; as a result, it will not be taken seriously by it as a political force. It remains outside of what Chomsky labels the realm of '"responsible" criticism', and instead occupies the position of '"sentimental", or "emotional", or "hysterical" criticism', and is excluded from debate because it transgresses the boundaries of accepted discussion (Chomsky 1967). Yet such criticism has an effect, albeit not in terms of generating tighter government policy. The infiltration of CAAT is one such indicator. More generally, Mike Turner, former Chief Executive of BAE Systems, complained that 'The thing that was always there at BAE was the campaign by people who frankly just didn't like the defence industry...It's gone on for many years but it is absolutely clear that the company has done nothing unlawful' (quoted in CAAT 2009d, 10). Such activity does not lead to tighter governmental policy, but it does damage the reputation of arms companies and act as an irritant. There remains, however, the challenge of how to translate this into practical change. The imposition of costs on the state and industry – through legal challenges, protest and embarrassment through media work and direct action – is already an intervention in the operation of the arms trade, but there are many, further steps to be taken before the orientation of UK policy, let alone the wider international arms trade, can be said to have shifted significantly.

The prospect of insiders undermining outsiders raises the question, however, of whether the social forces exist for CAAT's and BASIC's more transformist arguments to have force. Within CAAT, it is acknowledged that it has been harder to generate interest in the relationship between companies and government than in campaigns focusing on human rights abusers in the South. And BASIC's strategy is pitched at elites rather than a broader public. However, the work of CAAT and BASIC can be understood as complementary elements of the wider peace movement: BASIC uses an insider strategy to try to promote the vision of organisations such as CAAT and CND. Yet in the post-Cold War era, 'Peace movements of the old kinds are no longer viable' as the 'nuclear-pacifist and anti-imperialist paradigms' of Cold War social movements 'are no longer adequate for mobilising widespread responses' (Shaw 1994, 664). They have been displaced by the depoliticised campaigning of the human rights and development movements to which Amnesty and Oxfam are central. There thus remains a significant challenge in terms of the mobilisation of socially grounded, transformist responses to organised violence.

Reproducing hierarchy: global civil society and North–South relations

NGO activity on the arms trade seems at first glance to be a good example of the increasingly global nature of civil society. The Control Arms campaign, in particular, is self-consciously global in character, having so far gathered 1 million signatures from around the world in support of a treaty to regulate the arms trade. NGO partnerships with Southern organisations, either through the IANSA network, or through Saferworld and International Alert's programming and capacity building work in the South, are attempts to raise the voices of Southern populations and activists in pursuit of tougher regulation of the arms trade and improvements in human security. However, the uneven internationalisation of NGOs as actors, combined with ongoing Northern leadership of international campaigns, means that global civil society remains nationally circumscribed and Northern-dominated.

For example, it is the UK-based headquarters of Oxfam, Amnesty and IANSA, the three organisations leading the Control Arms campaign, that give campaign direction and make the major research contribution. More generally, Amnesty's 'transnational appearance [disguises] various national economic and social structures that significantly constrain its global aspirations' and it remains 'a surprisingly masculine (culturally), white, Western, and middle-class organisation' (Hopgood 2006, x–xi). Oxfam GB, meanwhile, 'dominates the coalition to the extent that, in practice, its coalition partners are subsidiaries'; Oxfam GB employees are the authors of the majority of Oxfam International policy documents, and most of Oxfam's funding and volunteers come from the UK (Berry and Gabay 2009, 347). CAAT's aims and its own international outreach similarly reproduce a national frame of reference, in that its activity is directed at the UK state and ENAAT participants are nationally based peace groups. And IANSA, the 'global' movement against gun violence, still reproduces a national-territorial cartography by grouping members according to nation-state and region, and it sees its role as facilitating locally based NGOs' relationships with their national states. However, such a national framing may in fact remain appropriate, all the talk of globalisation notwithstanding. That is, much as military production is internationalising, states are the arms industry's main customers, regulating the industry at the national level and providing the resources for its success. Conceptual disputes about the international, transnational or global nature of civil society discussed in Chapter One can thus be seen playing out in practice.

This is not to deny the existence or strength of Southern groups campaigning on small arms issues. What is notable, however, is the predominant focus on armed violence and small arms, rather than on the transfer of major conventional weaponry. In Tanzania, for example, activism on small arms is much more pronounced than in relation to the controversial air traffic control system deal of 2001. There are also significant regional disparities within Southern activism, as discussed in Chapter Six: Central and South Asia, and the Middle East have notably lower levels of activism than Sub-Saharan Africa. According to NGOers, this is due to technical issues such as the lack of native language speakers or resources, but also to the dominance of state-based definitions of security. But the language of civil society being lacking, or weak, in the South assumes that the European experience that created the structural preconditions for liberal democratic forms of political community and action can be universalised. And the social and political conditions for outward-oriented conceptions of security, which form the basis of dominant understandings of international relations, do not operate in most of the non-European world in the same way (Krause 1996, 320). Thus, while there is civil society activism on arms issues in the South, it formulates the problem differently to Northern organisations. This raises a host of difficulties for NGO attempts to generate global collective action on the arms trade. Indeed, it is not clear that it is the arms trade, *per se,* that is the problem. While weapons are used in a variety of forms of organised and individual violence, the focus on controlling the technologies of violence common to them may in fact obscure the different social relations of violence around the world.

The relative weakness of US-based pro-control NGOs is a notable feature of activism in relation to the arms trade, despite a more general Northern dominance of the NGO world. This is another way in which the liberal assumption of global civil society as a progressive sphere needs rethinking. If we understand the NRA and also arms capital to be members of global civil society, then US dominance of global civil society is more obvious. It is UK and European NGOs that have been most active on arms control, while US NGOs have been less able to influence the debate because of the strength of both the pro-gun domestic lobby and the military preponderance of the US state. US-based arms control NGOs, such as Center for Defense Information, Federation of American Scientists, and the Friends Committee on National Legislation, have not made much headway in US arms control debates, much less the international scene (Erickson 2007, 8). The ideological terrain of arms control is very different in the US, with the NRA acting as a powerful anti-Arms Trade Treaty voice.

In addition to reproducing a nationally based frame of reference, NGOs reproduce hierarchical relations in more subtle forms. Organisations taking the prefix Safer- have formed in the South, for example, Safer Albania, SaferAfrica, and Safer Rwanda, although none of these have any official connection to the London-based organisation that, in its name, takes responsibility for the whole world. There is a sense, articulated by one NGO staffer, that they are trading on Saferworld's reputation, attempting to create the impression of a direct link through its name. This is not to say that NGOs are unaware of such North–South hierarchies. They explicitly try to mitigate Northern dominance through, in IANSA's case for example, stipulating that neither the secretariat nor individual members have the mandate to represent or speak on behalf of IANSA in policy debates. This can be understood as an attempt to create a space for Southern voices to be heard on their own terms in international fora. International Alert, Oxfam and Saferworld have partnerships with Southern NGOs, and their training is largely understood as a means of facilitating Southern civil society participation in public life and international affairs. However, the intentions of NGOs notwithstanding, Chapter Six demonstrated the internal disputes over North–South relations within the IANSA network, and further research is needed to assess how non-governmental activism on arms issues is understood from a variety of points within the South. Moreover, educational metaphors remain a key feature of both classical and contemporary liberal theory and practice (Mehta 1999; Parekh 1995). This is because 'remedial development is not only a moral right, but can be justified as a form of enlightened self-interest' in contemporary development and security practice (Duffield 2001, 37), a statement that echoes the classical liberal desire to transform unfamiliar societies and cultures into liberal ones and the presentation of this as of universal benefit.

Marginalisation of questions of violence

The final key element of liberal accounts of global civil society identified in Chapter One is a strong emphasis on non-violence, both analytically and politically. One effect of this is to marginalise violence to the fringes of international relations, rather than understand it as playing a central role in the historical and contemporary development of capitalism and state formation. This naturalises the violence of the state and arms capital and means that debates about the arms trade start from the position of accepting the status quo, which brackets the violent history of international relations. While NGOs and civil society actors are expected to be non-violent, the state and capital are not. The emphasis on non-violence in global civil society usually rests on the acceptance of common definitions of violence as including violence against property (such as weapons themselves, and the headquarters of arms-producing companies). These definitions are thus infused with the spirit of capitalist social relations. The debate about non-violence 'should not, from the beginning, be clouded by ideologies which serve the perpetuation of violence' (Marcuse 1969, 116). That is, moral or strategic arguments about the use of violence by global civil society actors should not start from the premise that non-state violence is illegitimate, which leaves the violence carried out and sanctioned by the state and capital unchallenged. Yet the task of opening up debate about the nature and role of violence in social life and the use of force as a tool of activism, are marginalised by the emphasis on the non-violent character of global civil society.

This conceptual preference is borne out in empirical practice: even the most outsider of the organisations analysed in this book, CAAT, is explicit about its adherence to a code of non-violence that tends towards the more pacifist end of the spectrum even though, as an organisation, CAAT does not self-identify as pacifist. While non-violence is a key principle for formal organisations, more loosely organised groupings of activists are not always bound by the same logic. There is a tension between CAAT and direct action groups that do not eschew the use of physical force. While CAAT is seen as 'over the top' by the more insider NGOs, it is criticised by some direct action activists for not being outsider enough, for not taking a more overtly direct action stance. For example, CAAT seeks police approval for protests such as that against the Defence Systems and Equipment International (DSEi), a biennial defence and security exhibition/arms fair. Disarm DSEi, a direct action campaign targeting the exhibition, in contrast, does not negotiate with the police on principle. In light of calls for greater communication between police and activists in the aftermath of the G20 protests in London in early 2009, in advance of that year's exhibition, Disarm DSEi justified its stance in an open letter to New Scotland Yard. Amongst the reasons set out were that 'DSEi policing has been violent, intimidating, and repressive', including the use of searches under Section 44 of the Terrorism Act, arbitrary arrests, harassment and kettling (a policing tactic that involves cordoning off and containing protestors in an enclosed area, often for long periods of time), which 'has led us to mistrust the police'. In their view, the civil liberty of the right to protest is under threat in the UK, to the extent that the UK government and police force's role in 'ensuring that the arms dealers reach their destinations and their investors are not embarrassed or inconvenienced will always come before allowing public dissent' (Disarm DSEi 2009). The claim that it is the coercive arm of the state that is the violent actor is common to direct action activism, and is an attempt to reorient the debate around violence. While recent events such as the death of Ian Tomlinson after being struck by a police officer during the G20 protests in London in April 2009 (BBC 2009b), and the use of acoustic weapons against protestors in Pittsburgh at the September 2009 G20 summit (Weaver 2009), may be delegitimising Northern security forces' claim to neutrality, the marginalisation of CAAT on the basis of its protest activity and its need to be *seen* as non-violent signals that pro-state and pro-capital definitions of violence retain significant ideological weight.

Perpetuation of a hierarchical world military order

Overall, NGO interventions on arms trade issues play a key role in perpetuating a profoundly hierarchical world military order. That is, for the NGOs, Northern military spending and production is not a security problem in and of itself, although BASIC and CAAT pay lip service to this. The issues the NGOs raise in relation to it are largely articulated as defence-industrial. The main problem of the arms trade, according to the NGO community, is its impact, in particular the impact of small arms, on human rights, development and conflict in the South. This is a highly partial argument that reproduces the dominant conflict–security–development nexus that has become central to post-Cold War international security, severs small arms from the wider arms dynamic and world military order, and hives off violent conflict in the South from the wider spectrum of war and organised violence.

This construction of the problem posed by the arms trade rests on a Eurocentric historical understanding. That is, the historical experience of state formation in the North has created internally pacified states with outward-facing militaries (Giddens 1985). States can maintain high levels of military capability without their economies as a whole being dominated by military spending (Shaw 1991, 46). As such, militarism in its most common understandings, as either military build-up, military dominance of the economy, or the glorification of war, is less obvious within the North. A definition of militarism as the relationship between war preparation and society (Shaw 1991, 11), in contrast, allows for analysis of Northern militarism, including its differences from, and relationship to, Southern forms of militarism.

Within the North, war preparation has changed significantly through the increasing lethality of military equipment, especially in the form of nuclear weapons, which 'modified the ground rules of the world military order and the way in which war and war-preparation affected societies'. In the process, Northern societies themselves have accordingly become increasingly demilitarised (Shaw 1991, 13, 23). Militarisation is thus less apparent or obvious in the North, and inter-state relations within the North are less likely to be settled by the use of force. However, Northern states are still likely to use weaponry on Southern populations. Yet because Northern states can produce a greater proportion of their military equipment domestically, the use of military equipment against Southern populations tends not to be discussed as an arms trade issue. In addition, the advent of nuclear weapons and Cold War in the North contributed to the emergence of hot wars in the South. Through the 'enforced nuclear peace', the South 'took on central importance as site of armed conflict', with local proxy and client forces playing a pivotal role (Barkawi 2001, 110). Not only do Northern states use military equipment against Southern populations in direct attacks, but they also arm and train Southern military forces as part of the transnational constitution of force.

Processes of state formation and militarisation have taken a different form in the South, and cannot adequately be understood through the lens of the European historical experience. The role of colonial powers and, since decolonisation, external patrons, has been central to the shaping of the institutional and material basis of the state in the South. Arms sales have been an important way of building these relationships. The arms trade has historically played a role in the spread of the capitalist system into the periphery and the incorporation of non-arms-producing states into the world military order (Albrecht and Kaldor 1979). The emergence of industrial armies in the South was associated with industrialisation, the rise of urban elites, the spread of multinational manufacturing capital, and the development of authoritarian forms of rule (Albrecht and Kaldor 1979, 12-13). Military and military-related technology transfers have played a role in structuring patterns of wider technological development in the South. The importation of capital-intensive technology and the establishment of arms production capabilities increased Southern dependence on Northern suppliers, and shaped wider patterns of development (Lock and Wulf 1979, 211, 226). Arms purchases and the privileging of the role of the military have thus historically provided a coercive backbone to state apparatuses, protecting elites against potentially restive publics and providing the stability and predictability necessary for international capital to operate.

Coercion has been central to state formation: war and state-making are interdependent processes, with the organisation of a monopoly on violence a key task for states (Tilly 1985, 170-1). War preparation pre-dates the capitalist state system and is relatively autonomous from it, as is militarism as a phenomenon. It is thus important to recognise the specificity of warfare and military production (Barkawi 2006; Giddens 1985; Shaw 1988). The world military order has its own structures and characteristics that cannot simply be read off from capitalist social relations or the international states system. However, capitalism industrialised militarism and made it more lethal than ever before (Mann 1988), and military production straddles the realms of security and political economy. The development of capitalism has been reliant on military force to create the conditions for and underpin the spread of capital, in particular expropriation and the defence of private property (Mann 1988, 144). In these processes, the emergence of a relatively pacified core and violent periphery are interrelated. The extraction of surplus from the periphery and its redistribution in the core through imperialism functioned to 'ameliorate in the advanced countries social contradictions germane to capital accumulation' and provide 'the social conditions for relatively stable polyarchic political systems' based on consensual domination (Robinson 1996, 347).

While this has led to relatively stable social relations in the North, social life is more turbulent in the South as capital accumulation is more fragile and coercive, and authoritarian political forms or popular revolution have resulted (Robinson 1996, 360). As Mann puts it, 'capitalism has contained an institutionalised, relatively non-coercive core and an expropriated, militaristic periphery' (1988, 138). The development of authoritarian political systems in the South – those states more likely to be directly physically repressive using the means of violence obtained through the arms trade against their own populations, spending greater proportions of their resources on the military, that is, the states that NGOs are most concerned about – is thus related to the historical development of capitalism.

There has been a general process of the expansion and centralisation of the state apparatus in the South which, combined with frustrated socio-economic development, often led to opposition forces turning to the military, making post-independence military-backed coups commonplace (Bromley 1994, 97). This is part of an explanation of why Southern states are more likely to have authoritarian, repressive governments: there are structural pre-conditions for liberal democratic forms of political community that have not been instantiated in much of the South. That is, liberal forms of sovereign polity or liberal democratic systems of representation cannot emerge where pre-capitalist formations and/or the state are directly involved in organising material reproduction of society and appropriation of surpluses. This pattern of development provided the social base for authoritarian rule in much of the post-colonial world. And the military often played a role in trying to consolidate capitalist forms of production and rule and establish room for independent manoeuvre in the international system (Bromley 1994, 103, 106).

Many Southern states thus do not match up to the widespread definition of the modern state as a 'bordered power container' with a monopoly on violence, featuring internally pacified societies and the extrusion of violence into externally oriented aggression, represented institutionally in outward-facing militaries and inward-facing police forces (Giddens 1985, 120). With the rise of industrial capitalism 'a fundamental shift took place in the relationship between warfare, states and societies', in that internal pacification or demilitarisation 'created at the same time a capacity for greater violence by states against one another'. There is thus 'a leap in the levels of violence possible in a more consistently "outward-pointing" militarism' (Shaw 1991, 18; also Giddens 1985, 192). Such arguments often draw heavily on the social theory of Max Weber, from whom the idea of the state as territorially bound and exercising a monopoly on legitimate violence is derived. However, Weber only intended his definition of the state to apply to the European state of the late 19th and early 20th centuries, when he was writing, but his idea has since been appropriated as universal (Barkawi 2006, 43). So our dominant ideas about the character and role of the state are derived from the European world but applied universally. Thus, we cannot understand it when Southern states use violence against their own populations; when the distinction between the police and military is not so clear-cut; when arms transfers are used for purposes they would not normally be used for in Northern states, although the racialised and class-based exercise of state-sanctioned violence within the North mitigates even this claim. This leads commentators to castigate Southern states for being repressive, corrupt and so on. Without a historically sensitive account of the processes of state formation, solutions based on this understanding will continue to be inadequate.

Internationally agreed standards for the use of force are based on the European model of the state. Solutions are based on the ideal of an outward-facing military complementing an inward-facing police, states with a legitimate monopoly on violence, and world politics as inter-state relations. But juridical sovereignty should not be conflated with substantive sovereignty: just because governments around the world have signed up to these standards does not mean that they are institutionally in a position to carry them out or necessarily interested in doing so. The political economy of military development – in which Southern states are unable to produce weaponry across the board and are thus more dependent on imports – significantly qualifies the supposedly universal fact of state sovereignty and non-interference in domestic affairs. NGOs are thus attempting to reproduce substantive practices that are grounded in European historical experience. They are effectively asking Southern states to act as if history does not matter. They disconnect Southern practices from their historical and political context, and remove Northern practices as part of the wider problem. The rise of the good governance agenda in military spending, the focus on the opportunity costs of military spending and so on mean that Southern states' procurement practices are on the table because they have to import a greater proportion of their military equipment, while those of states that can afford to produce much of their own weaponry, or are protected by major arms suppliers, are not. Thus the North is deemed to be excluded from the need to change, which means that the problem becomes seen as internal to the South

– creating the South as a site of intervention and facilitating inappropriate practices.

Ramifications for NGO practice

The critique of NGO practice put forward in this book raises the question of what NGOs could do differently and what scholarship can offer in pursuit of change. The arms trade is facilitated by the interrelation of three hegemonic social formations: capitalism, militarism and imperialism. Therefore, the question for NGOs is, how can they best chip away at these formations through their activism on the arms trade? Their activism, whether aimed at regulation, reform or abolition of the arms trade, needs to be based on an understanding of the social forces that generate and perpetuate it. For example, the internationalisation of arms capital needs to be mapped in its relation to states in order better to understand the policies and practices of the state-sanctioned trade. Activism in relation to the state–capital relationship needs to be both grounded in national context and also internationalised. Northern and Southern forms of militarisation need to be assessed in a single analytical frame so that the historical links between Northern and Southern (in)security can be understood. And the North–South relations produced and reproduced by both the arms trade itself and activism in relation to it need to be analysed with a view to creating space for counter-hegemonic Southern initiatives to emerge, which can be supported in solidarity by Northern social movements and NGOs.

In terms of strategy, the broad lesson from this analysis is that insiders have the legitimacy that could allow them to wield more stick than carrot, be more openly critical of states and capital in their advocacy, policy programming and campaign work. However, as Hopgood has argued in relation to Amnesty International but with wider applicability, while it has the moral authority to 'name and shame', 'once one is off the fence it is hard to get back on' (Hopgood 2006, 204). In addition, a key feature of the work of an insider organisation such as BASIC is precisely based on a low-key, technical advocacy approach, backed up by the open criticism of groups such as CAAT. So it is not simply a question of strategy, but also one of argument and objective. Outsiders, meanwhile, could be more explicit about the interim steps that would lead to their transformative vision, and the ways in which the relationship between arms capital and the state are connected to the forms of Southern insecurity that are predominant in public understandings of the arms trade. Joining the dots between the problem as CAAT understands it and the problem as insiders represent it would possibly allow the public, MPs and policymakers to understand better the relevance of CAAT's campaigning.

However, it is important not to lapse into a pluralist idealism that assumes NGOs can necessarily all work collaboratively and make better strategic choices that lead to better regulation of the arms trade, let alone its reduction or even abolition. Civil society is a realm of contradiction and contestation, marked by the formal separation of public and private, the disparity in resources between reformists and transformists due to their relative relations to the state, capital and philanthropic organisations, and North–South hierarchies that mean non-state public action takes different forms around the world. We should also avoid overstating the internal homogeneity of NGOs: the NGO community is a small one within which there is a significant circulation of staff, and staff often move on to work for government and IGOs. As such, while there is significant evidence of the shared policy agenda between reformist NGOs, states and IGOs, the unexpected everyday agency of more radical individuals is not to be underestimated.

The task of this book has been to map NGO activism on the arms trade, highlighting the assumptions that drive NGO practice and subjecting them to critical scrutiny. Processes of translation can promote collaborative learning between scholars and practitioners, while accepting that they are engaged in socially distinct activities (Stavrianakis 2006, 153). Any concrete suggestions for change to emerge from this critique would need to be arrived at in dialogue with NGO staffers, or through NGO staffers adapting the arguments made here for their own purposes. Part of the argument has been to examine how and why NGOs operate in the way that they do, and with what effects. While NGO activism is conditioned in a variety of ways, NGOs are continually shaping and reworking the meaning of the arms trade, and there are always ways in which discourses and practices can be intervened in differently. Hegemony is never complete or final, and there are always avenues to be explored and weaknesses to be exploited.

Bibliography

Albrecht, Ulrich and Mary Kaldor (1979) 'Introduction', in Mary Kaldor and Asbjorn Eide (eds.) *The World Military Order. The Impact of Military Technology on the Third World* (London: Macmillan), pp. 1-16.

Amicus (2004) 'Amicus Urge MPs to Support the UK Defence Industry', *PR Newswire*, 28 September 2004, http://www.prnewswire.co.uk/cgi/news/release?id=131032 (10 April 2009).

Amnesty International (1995) *Rwanda: Arming the perpetrators of the genocide*, June 1995 (London: Amnesty International).

Amnesty International (1997) 'Indonesia and East Timor: Arms and security transfers undermine human rights', 3 June 1997, http://web.amnesty.org/library/Index/ENGASA210391997?open&of=ENG-390 (27 November 2006).

Amnesty International (2000) 'Indonesia: EU ban on military and security exports to Jakarta must not be lifted, for now', 14 January 2000, http://web.amnesty.org/library/Index/ENGASA210042000?open&of=ENG-IDN (27 November 2006).

Amnesty International (2001) *Amnesty International Campaigning Manual* (London: Amnesty International).

Amnesty International (2003a) 'Iraq: People come first – Protect human rights during the current unrest. Amnesty International's 10-point appeal', 22 April 2003, http://www.amnesty.org/en/library/asset/MDE14/089/2003/en/7d6aabb3-d6ff-11dd-b0cc-1f0860013475/mde 140892003en.pdf (7 October 2009).

Amnesty International (2003b) 'Iraq: Civilians under fire', April 2003, http://www.amnesty.org/en/library/asset/MDE14/071/2003/en/9ffda314-d709-11dd-b0cc-1f0860013475/mde140712003en.pdf (18 October 2009).

Amnesty International (2004) *Sudan. Arming the perpetrators of grave abuses in Darfur* (London: Amnesty International).

Amnesty International (2006) *Dead on Time. Arms transportation, brokering and the threat to human rights* (London: Amnesty International).

Amnesty International (2007) *A Global Arms Trade Treaty: What states want*, 17 October 2007, http://www.amnesty.org/en/library/asset/POL34/ 004/2007/ en/dom-POL340042007en.pdf (18 October 2009).

Amnesty International (2008) *Blood at the Crossroads: Making the case for a global Arms Trade Treaty*, September 2008 (London: Amnesty International).

Amnesty International (2009a) *Israel–OPT: Fuelling Conflict: Foreign arms supplies to Israel/Gaza*, February 2009 (London: Amnesty International).

Amnesty International (2009b) *Stopping the Terror Trade. How human rights rules in an arms trade treaty can help deliver real security*, October 2009 (London: Amnesty International).

Amnesty International (no date) 'The History of Amnesty International', http://www.amnesty.org/en/who-we-are/history (4 July 2008).

Amnesty International UK (1999) 'East Timor: UK arms moratorium needed now', 7 January 1999, http://www.amnesty.org.uk/news_details.asp?NewsID=13133 (24 May 2006).

Amnesty International UK (2000) 'Urgent Need for UK Arms Export Legislation – Government's human rights-centred foreign policy being undermined', 20 January 2000, http://www.amnesty.org.uk/deliver/document/14093.html (10 May 2006).

Amnesty International UK (2005a) 'Amnesty International UK's Finances 2004-5', http://www.amnesty.org.uk/give/aiukfinances/charts.shtml (13 February 2006).

Amnesty International UK (2005b) 'New Report Exposes Arms Exports from UK and other G8 Nations Fuelling Poverty and Human Rights Abuses', 22 June 2005, http://www.amnesty.org.uk/news_details.asp?NewsID=16184 (24 May 2006).

Amnesty International UK (no date) 'AIUK's Finances', http://www.amnesty.org.uk/give/aiukfinances/index.shtml (24 February 2006).

Amos, Valerie, Jack Straw and Adam Ingram (2003) 'Ministerial Foreword', in DfID, *Strengthening International Export Controls of Small Arms and Light Weapons. Implementing the UN Programme of Action*, Lancaster House, London, 14–15 January 2003 (London: DfID).

Anders, Holger (2005) 'NGOs and the Shaping of European Controls on Small Arms Exports', in Elke Krahmann (ed.) *New Threats and New Actors in International Security* (Houndmills: Palgrave Macmillan), pp. 177-96.

Anheier, Helmut, Marlies Glasius and Mary Kaldor (2001) 'Introducing Global Civil Society', in *ibid.* (eds.) *Global Civil Society 2001* (Oxford: Oxford University Press), pp. 3-22.

Anheier, Helmut and Sally Stares (2002) 'Records of Global Civil Society', in Marlies Glasius, Mary Kaldor and Helmut Anheier (eds.) *Global Civil Society 2002* (Oxford: Oxford University Press), pp. 241-377.

Arnove, Robert and Nadine Pinede (2007) 'Revisiting the "Big Three" Foundations', *Critical Sociology*, 33(3): 389-425.

Arquilla, John and David Ronfeldt (eds.) (1997) *In Athena's Camp. Preparing for Conflict in the Information Age* (Santa Monica: RAND Corporation).

BAE Systems (2008a) *BAE Systems Annual Report 2007. Delivering Global Growth*, http://www.investis.com/investors/downloads/annualreport2007.pdf (2 December 2009).

BAE Systems (2008b) 'Stakeholder Engagement', *Corporate Responsibility Report 2008*, http://www.baesystems.com/BAEProd/groups/public/documents/bae_publication/bae_pdf_cr08_stake_engage.pdf (26 November 2009).

BAE Systems (no date) 'Digital Media', http://production.investis.com/ukadvantage/campaign/digital/ (2 October 2009).

Barkawi, Tarak (1998) 'Strategy as Vocation: Weber, Morgenthau and Modern Strategic Studies', *Review of International Studies*, 24(2): 159-84.

Barkawi, Tarak (2001) 'War Inside the Free World: The democratic peace and the Cold War in the Third World', in Tarak Barkawi and Mark Laffey (eds.)

Democracy, Liberalism and War: Rethinking the democratic peace debate (Boulder: Lynne Rienner), pp. 107-28.

Barkawi, Tarak (2006) *Globalization and War* (Lanham: Rowman and Littlefield).

Barkawi, Tarak and Mark Laffey (2002) 'Retrieving the Imperial: Empire and International Relations', *Millennium*, 31(1): 109-27.

Barkawi, Tarak and Mark Laffey (2006) 'The Postcolonial Moment in Security Studies', *Review of International Studies,* 32(2): 329-52.

Barker, Alex and George Parker (2008) 'Fox Pledges to Realign Defence Policy', *Financial Times,* 17 June 2008.

Bartelson, Jens (2006) 'Making Sense of Global Civil Society', *European Journal of International Relations,* 12(3): 371-95.

BASIC (1999) 'ESDI: Right Debate, Wrong Conclusions', 4 August 1999, http://www.basicint.org/europe/ESDP/ESDIright_debate-1999.htm (10 August 2009).

BASIC (2000) 'A Conflict Prevention Service for the European Union', BASIC Research Report, 2 June 2000, http://www.basicint.org/pubs/Research/COPS.PDF (10 August 2009).

BASIC (2001a) 'BASIC Values and Distinctive Qualities', BASIC internal operational style memo.

BASIC (2001b) 'Memorandum from the British American Security Information Council (BASIC)', in Defence Committee, *First Report,* 14 February 2001, HC 115 (London: The Stationery Office).

BASIC (2001c) 'Blair Must Seek Middle Ground on Future EU Force', 20 February 2001, http://www.basicint.org/pubs/Press/2001feb-Blair_middle_ground.htm (10 August 2009).

BASIC (2001d) 'Past Practice – How BASIC works', BASIC internal operational document.

BASIC (2006a) 'Memorandum from British American Security Information Council (BASIC)', 7 March 2006, in House of Commons Defence Committee, *The Future of the UK's Strategic Nuclear Deterrent: The Strategic Context,* 20 June 2006 (London: The Stationery Office), Ev. 111-20.

BASIC (2006b) 'Memorandum from British American Security Information Council (BASIC)', 20 January 2006, in House of Commons Defence Committee, *The Defence Industrial Strategy,* 25 April 2006, Seventh Report of Session 2005-6 (London: The Stationery Office), Ev. 70-75.

BASIC (2007a) 'New Chair Takes Over for BASIC. New Co-Executive Director Appointed in London', 3 December 2007, http://www.basicint.org/pubs/Press/071203.htm (12 May 2008).

BASIC (2007b) 'A BASIC Success Story: Closure of DESO', http://www.basicint.org/update/report07.pdf (8 August 2009).

BASIC (2007c) 'Memorandum from British American Security Information Council (BASIC)', 15 January 2007, in House of Commons Defence Committee, *The Future of the UK's Strategic Nuclear Deterrent: the White Paper,* 27 February 2007 (London: The Stationery Office), Ev. 96-100.

BASIC (2007d) 'Memorandum from British American Security Information Council (BASIC)', 3 December 2007, House of Commons Defence Committee, *The Future of NATO and European Defence,* Ninth Report of Session 2007-8, 4 March 2008 (London: The Stationery Office), Ev. 123-136.

BASIC (no date, a) 'Transatlantic Security – Conventional Weapons', http://www. basicint.org/WT/wtindex.htm (6 April 2009).

BASIC (no date, b) 'About BASIC', http://www.basicint.org/about.htm (3 July 2008).

BASIC and Saferworld (2007) 'Memorandum from British American Security Information Council and Saferworld', 15 November 2007, in Defence Committee, *UK/US Defence Trade Cooperation Treaty*, Third Report of Session 2007–8, 4 December 2007 (London: The Stationery Office), Ev. 31-38.

BBC (2009a) 'BAE Systems Faces Bribery Charges', *BBC News*, 1 October 2009, http://news.bbc.co.uk/1/hi/8284073.stm (23 November 2009).

BBC (2009b) 'Footage Shows G20 Death Man Push', *BBC News*, 7 April 2009, http://news.bbc.co.uk/1/hi/england/london/7988828.stm (1 December 2009).

Bell, Louise (2003) *The Global Conflict Prevention Pool. A joint UK Government approach to reducing conflict*, August 2003 (London: FCO).

Berry, Craig and Clive Gabay (2009) 'Transnational Political Action and "Global Civil Society" in Practice: The case of Oxfam', *Global Networks*, 9(3): 339-58.

Berryman, J. (2000) 'Russia and the Illicit Arms Trade', *Crime, Law and Social Change*, 33(1-2): 85-104.

Bissell, William Cunningham (1999) 'Colonial Constructions: Historicizing Debates on Civil Society in Africa', in John Comaroff and Jean Comaroff (eds.) *Civil Society and the Political Imagination in Africa. Critical Perspectives* (Chicago: University of Chicago Press), pp. 124-59.

Biting the Bullet (2006) *Reviewing Action on Small Arms 2006. Assessing the First Five Years of the Programme of Action* (London/Bradford: International Alert/ Saferworld/University of Bradford).

Bitzinger, Richard A. (2003) *Towards a Brave New Arms Industry?* Adelphi Paper 356 (Oxford: Oxford University Press).

Blair, Tony (2002) 'Press Conference by Prime Minister Tony Blair', 20 June 2002, http://www.number10.gov.uk/Page2999 (22 June 2009).

Blakeley, Ruth (2009) *State Terrorism and Neoliberalism: The North in the South* (London: Routledge).

Blaney, David L. and Mustapha Kamal Pasha (1993) 'Civil Society and Democracy in the Third World: Ambiguities and possibilities', *Studies in Comparative International Development*, 28(1): 3-24.

Bob, Clifford (2007) 'Conservative Forces, Communications and Global Civil Society: Towards conflictive democracy', in Martin Albrow, Helmut Anheier, Marlies Glasius, Monroe Price, and Mary Kaldor (eds.) *Global Civil Society 2007/8: Communicative Power and Democracy* (London: Sage), 198-203.

Bondi, Loretta (2003) 'Loretta Bondi's response', *SAIS Review*, 23(1), Winter-Spring 2003, 322-4.

Booth, Ken (1994) 'Strategy', in A.J.R. Groom and Margot Light (eds.) *Contemporary International Relations: A Guide to Theory* (London: Frances Pinter), pp. 109-27.

Bourne, Mike, Malcolm Chalmers, Tim Heath, Nick Hooper and Mandy Turner (2004) *The Impact of Arms Transfers on Poverty and Development* (Bradford: Centre for International Cooperation and Security).

Boutros-Ghali, Boutros (1995) *Supplement to an Agenda for Peace: Position paper of the Secretary-General on the occasion of the fiftieth anniversary of the United Nations*, 3 January 1995, http://www.un.org/Docs/SG/agsupp.html (6 November 2006).

Boutwell, Jeffrey, Michael Klare and Laura W. Reed (eds.) (1995) *Lethal Commerce: The global trade in small arms and light weapons* (Cambridge, MA: American Academy of Arts and Sciences).

Boutwell, Jeffrey and Michael Klare (eds.) (1999) *Light Weapons and Civil Conflict: Controlling the tools of violence* (Lanham: Rowman and Littlefield).

Brem, Stefan and Ken Rutherford (2001) 'Walking together or divided agenda? Comparing landmines and small arms campaigns', *Security Dialogue*, 32(2): 169-86.

Broek, Martin and Wendela de Vries (2006) 'The Arms Industry and the European Constitution', European Network Against Arms Trade, January 2006, http://www.caat.org.uk/publications/government/ENAAT-EU-report_web.pdf (21 May 2008).

Bromley, Simon (1994) *Rethinking Middle East Politics. State formation and development* (Cambridge: Polity).

Brown, Gordon (2007) 'Machinery of Government: Defence Trade Promotion', written Ministerial Statement, *Hansard*, 25 July 2007, http://www.publications.parliament.uk/pa/cm200607/cmhansrd/cm070725/wmstext/70725m0005.htm#07072565000335 (18 October 2009).

Burnell, Peter (1998) 'Britain's New Government, New White Paper, New Aid?' *Third World Quarterly*, 19(4): 787-802.

Business and Enterprise, Defence, Foreign Affairs and International Development Committees (2008) *Scrutiny of Arms Export Control (2008): UK Strategic Export Controls Annual Report 2006, Quarterly Reports for 2007, licensing policy and review of export control legislation*, 17 July 2008, HC254.

Buzan, Barry and Eric Herring (1998) *The Arms Dynamic in World Politics* (Boulder: Lynne Rienner).

CAAT (1996) 'How to answer questions on the European Code of Conduct on the Arms Trade', *Scott Special. Briefing 2*, 13 February 1996.

CAAT (2000) 'Memorandum from the Campaign Against Arms Trade (30 November 2000)', in Defence Committee, *First Report*, 14 February 2001, HC 115 (London: The Stationery Office).

CAAT (2001) 'CAAT Issues Condemnation of Arms Fair', 6 September 2001, http://www.caat.org.uk/information/press.php?url=060901prs (21 November 2003).

CAAT (2002a) 'Arms Trade Economics: Subsidies factsheet', February 2002, http://www.caat.org.uk/resources/publications/economics/subsidies-factsheet-0202.php (15 June 2009).

CAAT (2002b) 'F-16 Upgrades to the Israeli Air Force', 9 July 2002, http://www.caat.org.uk/press/archive.php?url=120702prs (10 August 2009).

CAAT (2002c) 'Clean Investment Campaign–BAe Systems 2002', http://www.caat.org.uk/campaigns/clean-investment-campaign/baes-2002.php (22 November 2003).

CAAT (2004) 'Arms Trade Subsidies Factsheet', May 2004, http://www.caat.org.uk/resources/publications/economics/subsidies-factsheet-0504.php (15 June 2009).

CAAT (2005a) *Who Calls the Shots? How government–corporate collusion drives arms exports* (London: CAAT).

CAAT (2005b) '2005 CAAT Steering Committee statement on spying', http://www.caat.org.uk/about/spying.php (26 July 2009).

CAAT (2005c) 'Memorandum from the Campaign Against Arms Trade', in Defence Committee *et al.*, *Strategic Export Controls. HMG's Annual Report for 2003, Licensing Policy and Parliamentary Scrutiny*, HC145 (London: The Stationery Office), Ev. 55-8.

CAAT (2005d) 'The G8: Arms dealers to the world', http://www.caat.org.uk/publications/government/g8armsdealers.php (28 April 2009).

CAAT (2006a) *An Introduction to the Arms Trade*, August 2006, http://www.caat.org.uk/publications/intro-briefing-2006.pdf (20 May 2008).

CAAT (2006b) 'Memorandum from Campaign Against Arms Trade (CAAT)', in Defence Committee *et al.*, *Strategic Export Controls: Annual Report for 2004, Quarterly Reports for 2005, Licensing Policy and Parliamentary Scrutiny*, HC873 (London: The Stationery Office), Ev. 158-162.

CAAT (2006c) 'UK Arms Sales to Saudi Founded on Bribery. MoD misleading of Parliament exposed', June 2006, http://www.caat.org.uk/issues/saudi-bribery.php (17 July 2008).

CAAT (2007a) 'Bringing the Judicial Review', http://www.controlbae.org/jr/bringingjr.php (29 July 2009).

CAAT (2007b) 'Government Announces Closure of Arms Marketing Unit after Public Campaign', 26 July 2007, http://www.caat.org.uk/press/archive.php?url=260707prs (8 August 2009).

CAAT (2008a) 'Arms Trading at UKTI', November 2008, http://www.caat.org.uk/publications/government/UKTI_briefing.php#refs (16 April 2009).

CAAT (2008b) 'UKTI: Armed and dangerous', http://www.caat.org.uk/campaigns/ukti/ (28 November 2008).

CAAT (2008c) 'Memorandum from Campaign Against Arms Trade', in Business and Enterprise Committees *et al.*, *Scrutiny of Arms Export Control (2008): UK Strategic Export Controls Annual Report 2006, Quarterly Reports for 2007, licensing policy and review of export control legislation*, HC254, 17 July 2008 (London: The Stationery Office), Ev. 79-80.

CAAT (2009a) 'Arms Trade Jobs – An overview', 23 June 2009, http://www.caat.org.uk/issues/jobs/jobs_overview.php (8 August 2009).

CAAT (2009b) 'CAAT Responds to David Miliband's Statement on Arms Sales to Israel', 21 April 2009, http://www.caat.org.uk/press/archive.php?url=210409prs (19 September 2009).

CAAT (2009c) 'Saudi Arabia', http://www.caat.org.uk/issues/saudi-arabia.php (28 April 2009).

CAAT (2009d) *CAAT News*, Issue 213, July–September 2009 (London: CAAT).

CAAT (2010a) 'High Court grants injunction against BAE settlement', 2 March 2010, http://www.caat.org.uk/issues/bae/bae20100302prs.php (12 March 2010).

CAAT (2010b) 'Campaigners continue to raise concerns about SFO-BAE settlement process', 8 April 2010, http://www.caat.org.uk/press/recent.php?url=20100408prs (9 April 2010).

CAAT (no date, a) 'About CAAT', http://www.caat.org.uk/about/about.php (3 July 2008).

CAAT (no date, b) 'Call the Shots', http://www.caat.org.uk/campaigns/calltheshots/calltheshots.php (3 July 2008).

CAAT (no date, c) 'Fundraising', http://www.caat.org.uk/fundraising (1 July 2005).

CAAT (no date, d) 'Arms Trade Treaty', http://www.caat.org.uk/issues/att.php (3 July 2008).

CAAT (no date, e) 'Action guidelines', http://www.caat.org.uk/about/actionguidelines.php (6 April 2009).

CAAT (no date, f) 'Call the Shots. Shut DESO', http://www.caat.org.uk/campaigns/calltheshots/actionpage.php (28 November 2008).

CAAT (no date, g) 'Disturbing Developments in Europe', http://www.caat.org.uk/issues/european-conduct.php (10 August 2009).

CAAT (no date, h) 'Global Poverty', http://www.caat.org.uk/issues/poverty.php (28 April 2009).

CAAT (no date, i) 'Corruption', http://www.caat.org.uk/issues/corruption.php (17 July 2008).

CAAT (no date, j) 'Indonesia', http://www.caat.org.uk/issues/indonesia.php (16 May 2006).

CAAT (no date, k) 'Small Arms, Mass Killing', http://www.caat.org.uk/issues/smallarms.php (17 July 2006).

CAAT and Corner House (2007) 'Control BAE: Reopen the Saudi corruption inquiry', http://www.controlbae.org/ (28 April 2009).

Cameron, Maxwell A., Robert J. Lawson and Brian M. Tomlin (eds.) (1998) *To Walk Without Fear: The Global Movement to Ban Landmines* (New York: Oxford University Press).

Cammack, Paul (2004) 'What the World Bank Means by Poverty Reduction, and Why it Matters', *New Political Economy,* 9(2): 189-211.

Castells, Manuel (2008) 'The New Public Sphere: Global civil society, communication networks, and global governance', *Annals of the American Academy of Political and Social Science* (616): 78-93.

Cattaneo, Silvia and Keith Krause (2004) 'A Voice for Whom: Legitimacy, representation and advocacy in the international action network on small arms', paper presented at the International Studies Association Annual Convention, Montreal, March 2004.

Chakrabarty, Dipesh (2000) *Provincializing Europe. Postcolonial Thought and Historical Difference* (Princeton: Princeton University Press).

Chanaa, J. (2005) 'Arms Sales and Development: Making the critical connection', *Development in Practice,* 15(5): 710-16.

Chatterjee, P. (1990) 'A Response to Taylor's "Modes of Civil Society"', *Public Culture,* 3(1): 119-34.

Chipman, John (1992) 'The Future of Strategic Studies: Beyond even grand strategy', *Survival,* 34(1): 109-31.

Chomsky, Noam (1967) 'The Responsibility of Intellectuals', *New York Review of Books,* 23 February 1967, http://www.chomsky.info/articles/19670223.htm (22 January 2007).

Chomsky, Noam (2003) *Understanding Power: The Indispensable Noam Chomsky,* Peter R. Mitchell and John Schoeffel (eds.) (London: Vintage).

Chomsky, Noam and Edward S. Herman (1979) *The Washington Connection and Third World Fascism* (Cambridge, MA: South End Press).

Clarion Events (no date) 'Defence and Security Industry – Quotes', http://defence.clarionevents.com/?page=support (26 November 2009).

Clark, Ann Marie (2001) *Diplomacy of Conscience. Amnesty International and Changing*

Human Rights Norms (Princeton: Princeton University Press).

Clegg, Liz (1999) 'NGOs Take Aim', *Bulletin of the Atomic Scientists,* 55(1): pp. 49-51.

Clinton, Hillary Rodham (2009) 'US Support for the Arms Trade Treaty', 14 October 2009, http://www.state.gov/secretary/rm/2009a/10/130573.htm (15 November 2009).

Coe, Jim and Henry Smith (2003) *Action Against Small Arms. A Resource and Training Handbook* (London/Oxford: International Alert, Oxfam GB, Saferworld).

Cohen, Jean L. and Andrew Arato (1992) *Civil Society and Political Theory* (Cambridge, MA: MIT Press).

Colas, Alejandro (2002) *International Civil Society. Social Movements in World Politics* (Cambridge: Polity).

Control Arms (2003) *Shattered Lives. The Case for Tough Arms Control* (London and Oxford: Amnesty International and Oxfam).

Control Arms (2004a) *Guns or Growth? Assessing the Impact of Arms Sales on Sustainable Development,* June 2004 (London and Oxford: Amnesty International, IANSA and Oxfam International).

Control Arms (2004b) *Guns and Policing. Standards to Prevent Misuse,* February 2004 (London and Oxford: Amnesty International, IANSA and Oxfam International).

Control Arms (2005) *The G8: Global Arms Exporters. Failing to prevent irresponsible arms transfers,* Control Arms Briefing Paper, June 2005 (London and Oxford: Amnesty International, IANSA and Oxfam International).

Control Arms (2006a) 'Control Arms Campaign: UN General Assembly votes for historic Arms Trade Treaty proposal', 7 December 2006, http://www.controlarms.org/en/media/2006/7-december-2006-control-arms-campaign-un-general-1 (30 July 2009).

Control Arms (2006b) *Arms Without Borders,* Control Arms campaign report, October 2006, http://www.controlarms.org/documents/Arms%20Without%20Borders_Final_21Sept06.pdf (16 July 2008).

Control Arms (2008a) 'Frequently Asked Questions on the Arms Trade and the Arms Trade Treaty', April 2008, http://www.controlarms.org/en/documents%20and%20files/frequently-asked-questions-on-the-arms-trade-and (9 April 2009).

Control Arms (2008b) 'Briefing for House of Lords debate', 16 May 2008, http://www.saferworld.org.uk/images/pubdocs/Control%20Arms%20briefing%20May%2008.pdf (25 November 2009).

Control Arms (2009) 'World's Biggest Arms Traders Promise Global Arms Treaty', 30 October 2009, http://www.controlarms.org/en/media/index.htm (15 November 2009).

Cook, Robin (1997a) 'Mission Statement for the Foreign and Commonwealth Office', speech delivered at Locarno Suite, FCO, London, 12 May 1997, reproduced in *The Guardian,* 12 May 1997, http://www.guardian.co.uk/indonesia/Story/0,2763,190889,00.html (10 May 2006).

Cook, Robin (1997b) 'Human Rights into a New Century', speech at the FCO, London, 17 July 1997, www.fco.gov.uk/servlet/Front?pagename=OpenMarket/Xcelerate/ShowPage&c=Page&cid=1007029391647&a=KArticle&aid=1013618392902 (10 May 2006).

Cooper, Neil (2000) 'The Pariah Agenda and New Labour's Ethical Arms Sales Policy', in Richard Little and Mark Wickham-Jones (eds.) *New Labour's Foreign Policy. A New Moral Crusade?* (Manchester: Manchester University Press), pp. 147-67.

Cooper, Neil (2006a) 'Putting Disarmament Back in the Frame', *Review of International Studies*, 32: 353-76.

Cooper, Neil (2006b) 'What's the Points of Arms Transfer Controls?' *Contemporary Security Policy*, 27(1): 118-37.

Corner House (2008) 'Sustaining Corruption Legal Challenges in a Hostile Political Environment When Corruption Investigations Have Not Been Sustained: Insights from "the BAE case"', 1 November 2008, http://www.thecornerhouse.org.uk/pdf/document/IACC.pdf (24 September 2009).

Corner House and CAAT (2007) 'Documents Reveal that Blair Urged End to BAE–Saudi Corruption Investigation', 21 December 2007, http://www.thecornerhouse.org.uk/item.shtml?x=559591 (14 August 2009).

Council of the European Union (1998) *European Code of Conduct on Arms Exports*, 5 June 1998, http://www.consilium.europa.eu/uedocs/cmsUpload/08675r2en8.pdf (29 March 2010).

Council of the European Union (2008) *Council Common Position 2008/944/ CFSP Defining Common Rules Governing Control of Exports of Military Technology and Equipment*, 8 December 2008, http://eur-lex.europa.eu/LexUriServ/LexUriServ.do?uri=OJ:L:2008:335:0099:0103:EN:PDF (17 March 2010).

Cox, Robert W. (1983) 'Gramsci, Hegemony and International Relations: An essay in method', *Millennium*, 12(2): 162-75.

Cox, Robert (1999) 'Civil Society at the Turn of the Millennium: Prospects for an alternative world order', *Review of International Studies*, 25: 3-28.

Cramer, Christopher (2007) *Violence in Developing Countries. War, Memory, Progress* (Bloomington: Indiana University Press).

Darby, Phillip (2004) 'Pursuing the Political: A postcolonial rethinking of relations international', *Millennium* 33(1): 1-32.

Davis, Ian (2007) 'Four Major UK Security Announcements on Eve of Parliamentary Recess', 30 July 2007, BASIC Press Release, http://www.basicint.org/pubs/Press/070730.htm (12 May 2008).

Davis, Ian (2006) 'A Dangerous Illusion', *The Guardian*, 15 December 2006, http://www.guardian.co.uk/commentisfree/2006/dec/15/thesaudiarmsdealacancera (4 July 2008).

Davis, Ian and Roy Isbister (2003) 'EU and US Cooperation on Arms Export Controls in a Post-9/11 World. Report of a roundtable hosted by BASIC and Saferworld', 23 January 2003, http://www.basicint.org/pubs/Joint/EUUSemReport.pdf (17 March 2010).

D'Cunha, Beccie, Anna Jones and Stefan Luzi (2007) 'Shut 'Em Down', *CAAT News*, December 2006–January 2007 (London: CAAT).

De Sousa Santos, Boaventura (2005) 'The Future of the World Social Forum: The work of translation', *Development*, 48(2): 15-22.

Defence, Foreign Affairs, International Development and Trade and Industry Committees (2003) *Strategic Export Controls: Annual Report for 2001, Licensing Policy and Prior Parliamentary Scrutiny*, HC474 (London: The Stationery Office).

Defence Industries Council (2009) 'Defence Industry Launches Reports to

Reconnect with the Country', 1 September 2009, http://www.defencematters. co.uk/News-Featured-Article/DIC-reports.aspx (29 September 2009).

Defence Manufacturers' Association (2006) 'Arms Trade Treaty', *DMA News,* Issue 35, January 2006, http://www.the-dma.org.uk/Intro/Newsletters/78.PDF (13 June 2006).

Desmond, Cosmas (1983) *Persecution East and West: Human Rights, Political Prisoners and Amnesty* (Harmondsworth: Penguin).

DfID (2003) *Tackling Poverty by Reducing Armed Violence. Recommendations from a Wilton Park Workshop, 14–16 April 2003* (London: DfID), http://www.dfid. gov.uk/Documents/publications/tacklingpovredviolence.pdf (17 March 2010).

DfID (2007a) 'Global Conflict Prevention Joint Pool – Small arms and light weapons projects', Freedom of Information response to author, 29 October 2007.

DfID (2007b) 'Memorandum from the Department for International Development (DFID)', in Defence Committee *et al., Strategic Export Controls: 2007 Review,* HC117, Ev. 71-7.

DfID (no date) 'UK Policy and Strategic Priorities on Small Arms and Light Weapons 2004–2006', http://www.dfid.gov.uk/pubs/files/policysmallarmsweapons.pdf (28 July 2004).

DfID, FCO, MoD (2002) *Small Arms and Light Weapons: A Policy Briefing* (London: DfID).

Disarm DSEi (2009) 'We Will Not Negotiate – An open letter to the police from Disarm DSEi', 3 September 2009, http://www.dsei.org/we-will-not-negotiate-an-open-letter-to-the-police-from-disarm-dsei (5 October 2009).

Doty, Roxanne Lynne (1996) *Imperial Encounters: The Politics of Representation in North–South Relations* (Minneapolis: University of Minnesota Press).

Dover, Robert (2007a) *Europeanization of British Defence Policy* (Aldershot: Ashgate).

Dover, Robert (2007b) 'For Queen and Company: The role of intelligence in the UK's arms trade', *Political Studies,* 55(4): 683-708.

DSO [Defence and Security Organisation] (2009) 'About the UKTI Defence and Security Organisation', http://www.deso.mod.uk/about.htm (29 September 2009).

Duffield, Mark (2001) *Global Governance and the New Wars: The Merging of Development and Security* (London: Zed Books).

Duffield, Mark (2002a) 'Social Reconstruction and the Radicalization of Development: Aid as a relation of global liberal governance', *Development and Change,* 33(5): 1049-71.

Duffield, Mark (2002b) 'War as a Network Enterprise. The new security terrain and its implications', *Cultural Values,* 6(1 and 2): 153-65.

Duffield, Mark (2005) 'Getting Savages to Fight Barbarians: Development, security and the colonial present', *Conflict, Security and Development,* 5(2): 141-59.

Duffield, Mark (2007) *Development, Security and Unending War. Governing the World of Peoples* (Cambridge: Polity).

Dunne, J. Paul and Ron Smith (1992) 'Thatcherism and the UK defence industry', in Jonathan Michie (ed.) *The Economic Legacy 1979–1992* (London: Academic Press), pp. 91-111.

Dunne, J. Paul and Eamon Surry (2006) 'Arms Production', SIPRI (ed.), *SIPRI Yearbook 2006. Armaments, Disarmament and International Security* (Oxford: Oxford University Press), pp. 387-418.

Dyer, Susannah L. and Natalie J. Goldring (1996) 'Controlling Global Light Weapons Transfers: Working toward policy options', http://www.basicint.org/WT/plw/96-controlling_global.htm (28 July 2006).

Eavis, Paul (2002) 'SALW in the Horn of Africa and the Great Lakes Region: Challenges and ways forward', *The Brown Journal of World Affairs,* 9(1): 251-60.

Eisenhower, Dwight D. (1961) *Military-Industrial Complex Speech,* Public Papers of the Presidents, The Avalon Project at Yale Law School, http://www.yale.edu/lawweb/avalon/presiden/speeches/eisenhower001.htm (10 June 2005).

Ennals, Martin (1982) 'Amnesty International and Human Rights', in Peter Willetts (ed.) *Pressure Groups in the Global System* (London: Frances Pinter), pp. 63-83.

Erickson, Jennifer L. (2007) 'The Arms Trade Treaty. The Politics Behind the UN Process', European and Atlantic Security Research Unit Working Paper, Stiftung Wissenschaft und Politik, July 2007, http://www.swp-berlin.org/common/get_document.php?asset_id=4149 (24 June 2009).

Escobar, Arturo (1995) *Encountering Development. The Making and Unmaking of the Third World* (Princeton: Princeton University Press).

Etzioni, Amitai (2004) 'The Capabilities and Limits of the Global Civil Society', *Millennium,* 33(2): 341-53.

EU NGOs (2004) *Taking Control. The Case for a More Effective European Union Code of Conduct on Arms Exports,* September 2004, http://www.saferworld.org.uk/images/pubdocs/Taking%20control.pdf (12 August 2009).

EU NGOs (2008) 'EU NGO Submission to COARM on Harmonisation among EU Member States on End-use and Post-export Controls', May 2008, http://www.saferworld.org.uk/publications.php/316/eu_ngo_submission_to_coarm_on_harmonisation_among_eu_member_states_on_end_use_and_post_export_contro (19 November 2008).

European Defence Agency (2006) 'EU Ministers Welcome EDA Ideas to Increase Defence R&T Investment and Collaboration', 15 May 2006, http://www.eda.europa.eu/newsitem.aspx?id=41 (15 September 2009).

Eyre, Dana P. and Mark C. Suchman (1996) 'Status, Norms, and the Proliferation of Conventional Weapons: An institutional theory approach', in Peter J. Katzenstein, (ed.) *The Culture of National Security: Norms and Identity in World Politics* (New York: Columbia University Press), pp. 79-113.

FCO (2006) 'Conflict Prevention – Small Arms and Light Weapons', www.fco.gov.uk (28 June 2006).

FCO (2009) 'Conflict Funding', http://www.fco.gov.uk/en/about-the-fco/what-we-do/funding-programmes/conflict-funding/ (28 September 2009).

FCO (no date) 'Small Arms and Light Weapons – A Serious Global Problem', http://www.fco.gov.uk (28 July 2004).

Florini, Ann M. and P.J. Simmons (2000) 'What the World Needs Now?' in Ann M. Florini (ed.) (2000) *The Third Force. The Rise of Transnational Civil Society* (Tokyo/Washington, DC: Japan Center for International Exchange and Carnegie Endowment for International Peace), pp. 1-15.

Foley, Conor and Keir Starmer (1998) 'Foreign Policy, Human Rights and the United Kingdom', *Social Policy and Administration,* 32(5): 464-80.

Fox, Liam (2009) 'Radical Reform Needed at the MoD', speech given 7 September 2009, http://www.conservatives.com/News/Speeches/ 2009/09/Liam_Fox_Radical_reform_needed_at_the_MoD.aspx (24 September 2009).

Garcia, Denise (2006) *Small Arms and Security. New Emerging International Norms* (Abingdon: Routledge).

Garcia, Denise (2009) 'Arms Transfers Beyond the State-to-State Realm', *International Studies Perspectives,* 10: 151-68.

Garland, Elizabeth (1999) 'Developing Bushmen: Building Civil(ized) Society in the Kalahari and Beyond', in John Comaroff and Jean Comaroff (eds.) *Civil Society and the Political Imagination in Africa. Critical Perspectives* (Chicago: University of Chicago Press), pp. 72-103.

Gates, Robert M. (2009) Speech to Economic Club of Chicago, 16 July 2009, http://www.defenselink.mil/speeches/speech.aspx?speechid=1369 (23 November 2009).

Geneva Declaration (2008) *Global Burden of Armed Violence* (Geneva: Geneva Declaration Secretariat), http://www.genevadeclaration.org/fileadmin/docs/Global-Burden-of-Armed-Violence-full-report.pdf (6 December 2009).

Giddens, Anthony (1985) *The Nation-State and Violence. vol. 2 of A Contemporary Critique of Historical Materialism* (Cambridge: Polity).

Gilby, Nicholas (1999) 'Arms Exports to Indonesia', October 1999, http://www.caat.org.uk/publications/countries/indonesia-1099.php (24 April 2009).

Gilby, Nicholas (2001) 'Labour, Arms and Indonesia: Has anything changed?' July 2001, http://www.caat.org.uk/publications/countries/labour-indonesia-0701.php (16 May 2006).

Goldie, Mark (1993) 'Introduction', in John Locke, *Two Treatises of Government,* Mark Goldie (ed.) (London: Everyman).

Goldring, Natalie J. (2006) 'Two Sides of the Same Coin: Establishing Controls for SALW and Major Conventional Weapons', *Contemporary Security Policy,* 27(1): 85-99.

Gongora, Thierry (1997) 'War Making and State Power in the Contemporary Middle East', *International Journal of Middle East Studies,* 29(3): 323-40.

Gordon, Uri (2007) *Anarchy Alive! Anti-Authoritarian Politics from Practice to Theory* (London: Pluto Press).

Gow, David (2003) '5,000 Jobs Safe as India Buys Hawks', *The Guardian,* 4 September 2003.

Gramsci, Antonio (1971) *Selections from the Prison Notebooks of Antonio Gramsci,* translated and edited by Quintin Hoare and Geoffrey Nowell-Smith (London: Lawrence and Wishart).

Grant, Wyn (1978) *Insider Groups, Outsider Groups and Interest Group Strategies in Britain,* Working Paper 19, Department of Politics, University of Warwick.

Grant, Wyn (1989) *Pressure Groups, Politics and Democracy in Britain* (London: Philip Allan).

Grillot, Suzette R., Craig S. Stapley and Molly E. Hanna (2006) 'Assessing the Small Arms Movement: The Trials and Tribulations of a Transnational Network', *Contemporary Security Policy,* 27(1): 60-84.

Grimmett, Richard F. (2008) *Conventional Arms Transfers to Developing Nations, 2000-2007,* 23 October 2008, Congressional Research Service Report for Congress, http://www.fas.org/sgp/crs/weapons/RL34723.pdf (5 August 2009).

Grossberg, Lawrence (ed.) (1986) 'On Postmodernism and Articulation. An Interview with Stuart Hall', *Journal of Communication Inquiry,* 10(2): 45-60.

Hall, Stuart (1986) 'Gramsci's Relevance for the Study of Race and Ethnicity',

Journal of Communication Inquiry, 10(2): 5-27.

Hartley, Keith (2008) 'Collaboration and European Defence Industrial Policy', *Defence and Peace Economics*, 19(4): 305-15.

Hartung, William D. (2008) 'In the Wake of Wall Street Bailout, Stop Bailing Out the Arms Industry', New America Foundation, 15 October 2008, http://www.newamerica.net/publications/articles/2008/wake_wall_street_bailout_stop_bailing_out_arms_industry_8216 (6 December 2009).

Hearn, Julie (2001) 'The "Uses and Abuses" of Civil Society in Africa', *Review of African Political Economy* (28)87: 43-53.

Hearn, Julie (2007) 'African NGOs: The New Compradors?' *Development and Change*, 38(6): 1095-110.

Held, David (1995) *Democracy and the Global Order. From the Modern State to Cosmopolitan Governance* (Cambridge: Polity).

Held, David (2004) *Global Covenant. The Social Democratic Alternative to the Washington Consensus* (Cambridge: Polity).

Held, David and Anthony McGrew, David Goldblatt and Jonathan Perraton (1999) *Global Transformations: Politics, Economics and Culture* (Cambridge: Polity).

Hencke, David (2008) 'MoD Plans Raid on Landmine Removal Fund to Keep Tornados Flying in Iraq', *The Guardian*, 10 March 2008.

Hensel, Nayantara (2008) 'Globalization and the US Defense-Industrial Base: The competition for a new aerial refuelling tanker', *Business Economics*, 43(4): 45-56.

Hewitt, Patricia (2002) *Hansard*, Written Answers, Column 309W-311W, 26 September 2002, http://www.publications.parliament.uk/pa/cm200102/cmhansrd/vo020919/text/20919w75.htm#20919w75.html_spnew7 (17 March 2010).

Hewitt, Patricia (2003) 'Terrorism: The price we pay for poverty', *New Statesman*, 3 February 2003.

Hill, Stephen M. (2006) 'Introduction: Future directions in small arms control', *Contemporary Security Policy*, 27(1): 1-11.

Hobson, John (2004) *The Eastern Origins of Western Civilisation* (Cambridge: Cambridge University Press).

Holtom, Paul (2007) *Small Arms Production in Russia*, March 2007 (London: Saferworld).

Hope, Chris (2004) 'BAE Warns of All-Out War from Hoon', *The Daily Telegraph*, 3 May 2004.

Hopgood, Stephen (2000) 'Reading the Small Print in Global Civil Society: The inexorable hegemony of the liberal self', *Millennium*, 29(1): 1-25.

Hopgood, Stephen (2006) *Keepers of the Flame. Understanding Amnesty International* (Ithaca: Cornell University Press).

Howard, Michael (2005) 'Howard: Conservatives will invest in Britain's armed forces', speech delivered on 30 March 2005, reproduced at http://www.defense-aerospace.com/article-view/verbatim/54932/uk-tories-promise-increased-defense-spending.html (7 August 2009).

Hubert, Don (2000) 'The Landmine Ban: A case study in humanitarian advocacy', Occasional Paper no. 42, Watson Institute Humanitarianism and War Project.

Human Rights Watch (1995) *Rwanda/Zaire: Rearming with Impunity. International Support for the Perpetrators of the Rwandan Genocide* (New York: Human Rights Watch).

IANSA (no date, a) 'About Us', http://www.iansa.org/about.htm (25 July 2009).

IANSA (no date, b) 'International NGO Action Network on Small Arms. Founding Document of IANSA', http://www.iansa.org/about/Founddoc1.pdf (16 August 2009).

IANSA (no date, c) 'Key Issues', http://www.iansa.org/issues/index.htm (16 August 2009).

IANSA (no date, d) 'IANSA Members', http://www.iansa.org/about/members. htm (25 November 2009).

ICBL (International Campaign to Ban Landmines) (no date) 'Treaty Basics', http://www.icbl.org/index.php/icbl/Treaties/MBT/Treaty-Basics (14 November 2009).

Ingram, Paul (2006) Letter to *The Guardian*, 30 November 2006.

Ingram, Paul and Ian Davis (2001) *The Subsidy Trap. British Government Financial Support for Arms Exports and the Defence Industry*, July 2001, Saferworld/BASIC, http://www.saferworld.org.uk/images/pubdocs/pubsubsidy.pdf (21 July 2008).

Ingram, Paul and Roy Isbister (2004) *Escaping the Subsidy Trap. Why Arms Exports Are Bad for Britain*, BASIC, Saferworld, Oxford Research Group, http://www.basicint.org/pubs/subsidy.pdf (21 July 2008).

Insight (2003) 'How the Woman at No 27 Ran Spy Network for an Arms Firm', *Sunday Times*, 28 September 2003, http://www.timesonline.co.uk/tol/news/uk/article1163959.ece (4 July 2008).

International Alert (2006a) 'Small Arms and Light Weapons', February 2006, http://www.international-alert.org/our_work/themes/security_1.php (3 July 2008).

International Alert (2006b) 'Security', February 2006, http://www. international-alert.org/our_work/themes/security.php?page=work& ext=set (3 July 2008).

International Alert (2006c) 'Biting the Bullet', http://www.international-alert.org/our_work/themes/biting_the_bullet.php (3 July 2008).

International Alert (no date) 'About Us', http://www.international-alert.org/about_alert/index.php?page=about (19 February 2006).

IRIN (2006) *Guns Out of Control: The continuing threat of small arms*, IRIN In-Depth Report, May 2006, available at http://www.irinnews.org/pdf/in-depth/Small-Arms-IRIN-In-Depth.pdf (19 November 2009).

Jackson, Paul (2003) 'Warlords as Alternative Forms of Governance', *Small Wars and Insurgencies*, 14(2): 131-50.

Jahn, Beate (2005) 'Barbarian Thoughts: Imperialism in the philosophy of John Stuart Mill', *Review of International Studies*, 31(3): 599-618.

Jameson, Angela (2006) 'BAE Systems Chief Reaps Reward for Years of Fighting for Revival', *The Times*, 27 February 2006.

Jordan, A.G. and J.J. Richardson (1987) *Government and Pressure Groups in Britain* (Clarendon Press, Oxford).

Jordan, Tim (2002) *Activism! Direct Action, Hacktivism and the Future of Society* (London: Reaktion Books).

JRCT (Joseph Rowntree Charitable Trust) (2007) 'Peace Grants Policy', http://www.jrct.org.uk/text.asp?section=0001000200010003 (5 October 2009).

Joseph, Kate and Taina Susiluoto (2002) 'A Role for Verification and Monitoring in Small Arms Control?' *Verification Yearbook*, pp. 129-43.

Kaldor, Mary (1983) *The Baroque Arsenal* (London: Abacus).

Kaldor, Mary (1999) *New and Old Wars: Organised Violence in a Global Era* (Cambridge: Polity).

Kaldor, Mary (2003a) *Global Civil Society. An Answer to War* (Cambridge: Polity).
Kaldor, Mary (2003b) 'The Idea of Global Civil Society', *International Affairs*, 79(3): 583-93.
Kaldor, Mary, Helmut Anheier and Marlies Glasius (2005) 'Introduction', in Helmut Anheier, Marlies Glasius and Mary Kaldor (eds.) *Global Civil Society 2004/5* (London: Sage), pp. 1-25.
Kaldor, Mary and Asbjorn Eide (eds.) (1979) *The World Military Order. The Impact of Military Technology on the Third World* (Macmillan: London).
Kant, Immanuel [1795] (1991) 'Perpetual Peace: A philosophical sketch', in Hans Reiss (ed.) *Kant. Political Writings*, 2nd edition, trans. H.B. Nisbet (Cambridge: Cambridge University Press), pp. 93-130.
Karp, Aaron (2002) 'Laudable Failure', *SAIS Review*, 22(1): 177-93.
Karp, Aaron (2003) 'Aaron Karp's Response', *SAIS Review*, 23(1): 310-12.
Kates, Don B. (2003) 'Democide and Disarmament', *SAIS Review*, 23(1): 305-9.
Keane, John (2003) *Global Civil Society?* (Cambridge: Cambridge University Press).
Knighton, Ben (2003) 'The State as Raider among the Karamojong: "Where there are no guns, they use the threat of guns"', *Africa: Journal of the International African Institute*, 73(3): 427-55.
Kopel, David B. (2003) 'The UN Small Arms Conference', *SAIS Review*, 23(1): 319-22.
Krause, Keith (1994) 'Middle Eastern Arms Recipients in the Post-Cold War World', *Annals of the American Academy of Political and Social Science*, vol. 535, *The Arms Trade: Problems and Prospects in the Post-Cold War World*, September 1994, pp. 73-90.
Krause, Keith (1995) *Arms and the State: Patterns of military production and trade* (Cambridge: Cambridge University Press).
Krause, Keith (1996) 'Insecurity and State Formation in the Global Military Order: The Middle Eastern case', *European Journal of International Relations*, 2(3): 319-54.
Krause, Keith (2001) *Norm-Building in Security Spaces: The emergence of the light weapons problematic*, REGIS Working Paper (Quebec: Dépôt légal-Bibliothèque nationale du Canada).
Krause, Keith (2002) 'Multilateral Diplomacy, Norm Building, and UN Conferences: The case of small arms and light weapons', *Global Governance*, 8, 247-63.
Krause, Keith and Andrew Latham (1998) 'Constructing Non-Proliferation and Arms Control: The norms of Western practice', *Contemporary Security Policy*, Special Issue. *Culture and Security: Multilateralism, Arms Control and Security Building*, 19(1): 23-54.
Krishna, Sankaran (2001) 'Race, Amnesia, and the Education of International Relations', *Alternatives*, 26: 401-24.
Kurtz, Lester R. (1988) *The Nuclear Cage. A Sociology of the Arms Race* (Englewood Cliffs: New Jersey).
Laffey, Mark (2000) 'Locating Identity: Performativity, foreign policy and state action', *Review of International Studies*, 26: 429-44.
Laffey, Mark and Kathryn Dean (2002) 'A Flexible Marxism for Flexible Times: Globalization and historical materialism', in Mark Rupert and Hazel Smith (eds.) *Historical Materialism and Globalization* (London: Routledge), pp. 90-110.
Laipson, Ellen (ed.) (2006) 'Security Sector Reform in the Gulf', Henry L. Stimson

198 • TAKING AIM AT THE ARMS TRADE

Center, May 2006, http://www.stimson.org/swa/pdf/StimsonSSRGulf.pdf (22 January 2008).

Laurance, Edward and Rachel Stohl (2002) *Making Global Public Policy: The Case of Small Arms and Light Weapons*, Small Arms Survey Occasional Paper No. 7, December 2002 (Geneva: Small Arms Survey).

Leigh, David (2007) 'BAE's Secret $12m Payout in African Deal', *The Guardian*, 15 January 2007.

Leigh, David and Rob Evans (2006) '"National Interest" Halts Arms Corruption Inquiry', *The Guardian*, 15 December 2006.

Lens, Sidney (1970) *The Military-Industrial Complex* (London: Stanmore Press).

Liotta, P.H. (2002) 'Boomerang Effect: The convergence of national and human security', *Security Dialogue*, 33(4): 473-88.

Lipschutz, Ronnie D. (1992) 'Reconstructing World Politics: The emergence of global civil society', *Millennium*, 21(3): 389-420.

Lock, Peter and Herbert Wulf (1979) 'The Economic Consequences of the Transfer of Military-oriented Technology', in Mary Kaldor and Asbjorn Eide (eds.) *The World Military Order. The Impact of Military Technology on the Third World* (London: Macmillan), pp. 210-31.

Locke, John [1689] (1993) *Two Treatises of Government*, Mark Goldie (ed.) (London: Everyman).

Lubbers, Eveline and Wil van der Schans (2004) 'The Threat Response Spy Files. Or, a story about an arms manufacturer, a private intelligence company and many infiltrators', SpinWatch dossier, November 2004, http://www.evel.nl/spinwatch/TRReport.htm (29 July 2009).

Luckham, Robin (1984) 'Of Arms and Culture', *Current Research on Peace and Violence*, VII(1): 1-64.

Lumpe, Lore, Sarah Meek and R.T. Naylor (2000) 'Introduction to Gun-running', in Lumpe (ed.) *Running Guns. The Global Black Market in Small Arms* (London: Zed Books), pp. 1-12.

Mabee, Bryan (2009) *The Globalization of Security. State Power, Security Provision and Legitimacy* (Houndmills: Palgrave Macmillan).

Maloney, William A., Grant Jordan and Andrew M. McLaughlin (1994) 'Interest Groups and Public Policy: The Insider/Outsider Model Revisited', *Journal of Public Policy*, 14(1): 17-38.

Manji, Firoze and Carl O'Coill (2002) 'The Missionary Position: NGOs and development in Africa', *International Affairs*, 78(3): 567-83.

Mann, Michael (1988) *States, War and Capitalism* (Oxford: Blackwell).

Marcuse, Herbert (1969) 'Repressive Tolerance', in Robert Paul Wolff, Moore Barrington Jr., and Herbert Marcuse, *A Critique of Pure Tolerance* (London: Jonathan Cape), 93-187.

Marsh, David (1983) *Pressure Politics. Interest Groups in Britain* (London: Junction Books).

Marx, Karl [1843] (1975) 'On the Jewish Question', in Karl Marx, *Early Writings*, trans. Rodney Livingstone and Gregor Benton (New York: Vintage), pp. 211-41.

Mayhew, Emma (2005) 'A Dead Giveaway: A critical analysis of New Labour's rationales for supporting military exports', *Contemporary Security Policy*, 26(1): 62-83.

McNeill, William H. (1982) *The Pursuit of Power: Technology, Armed Force, and Society since AD 1000* (Chicago: University of Chicago Press).

Mehta, Uday Singh (1999) *Liberalism and Empire. A Study in Nineteenth-Century British Liberal Thought* (Chicago: University of Chicago Press).

Melman, Seymour (1970) *Pentagon Capitalism. The Political Economy of War* (New York: McGraw-Hill).

Mepham, David and Paul Eavis (2002) *The Missing Link in Labour's Foreign Policy. The Case for Tighter Controls over UK Arms Exports* (London: IPPR/Saferworld).

Miliband, David (2009) 'Israel (UK Strategic Export Controls)', written Ministerial Statements, *Hansard*, 21 April 2009, http://www.publications.parliament.uk/pa/cm/cmtoday/cmwms/archive/090421.htm#hddr_8 (15 June 2009).

Miller, Kathleen and Caroline Brooks (2001) 'Export Controls in the Framework Agreement Countries', BASIC Research Report, http://www.basicint.org/pubs/Research/2001ExportControls1.htm (21 May 2008).

Miller, Kathleen and Theresa Hitchens (2000) 'European Accord Threatens to Lower Export Controls', BASIC Paper, number 33, August 2000, http://www.basicint.org/pubs/Papers/BP33.htm (21 May 2008).

Mittelman, James H. (2005) 'Globalization, Cosmopolitanism, and the Kantian Revival: Commentary on David Held's "At the Global Crossroads"', *Globalizations*, 2(1): 114-16.

MoD (1998) *Strategic Defence Review. Modern Forces for the Modern World,* July 1998, Cm3999 (London: Ministry of Defence).

MoD (2002) *Defence Industrial Policy,* Ministry of Defence Policy Papers No. 5, October 2002, http://www.mod.uk/linked_files/issues/paper5/defence_Industrial.pdf (7 June 2005).

MoD (2005) *Defence Industrial Strategy. Defence White Paper* (London: The Stationery Office).

MoD (2007) 'Memorandum from the Ministry of Defence', in Business and Enterprise, Defence, Foreign Affairs and International Development Committees, *Scrutiny of Arms Export Control (2008): UK Strategic Export Controls Annual Report 2006, Quarterly Reports for 2007, licensing policy and review of export control legislation,* 3 July 2008, HC254 (London: The Stationery Office), Ev. 42-44.

Mohanty, Chandra Talpade (1997) 'Under Western Eyes: Feminist scholarship and colonial discourses', in Sandra Kemp and Judith Squires (eds.) *Feminisms* (Oxford: Oxford University Press), pp. 91-95.

Monbiot, George (2007) 'The Parallel Universe of BAE: Covert, dangerous and beyond the rule of law', *The Guardian*, 13 February 2007, http://www.guardian.co.uk/commentisfree/2007/feb/13/bae.foreignpolicy (4 July 2008).

Morris, Nigel and Stephen Khan (2005) 'Straw Pledges Curb on £15bn Arms Trade', *The Independent*, 16 March 2005.

Mouffe, Chantal (1998) 'Hegemony and New Political Subjects: Toward a new concept of democracy', Stanley Gray trans., in Cary Nelson and Lawrence Grossberg (eds.) *Marxism and the Interpretation of Culture* (Houndmills: Macmillan Education), pp. 89-101.

Muggah, Robert and Eric Berman (2001) 'Humanitarianism Under Threat: The humanitarian impacts of small arms and light weapons', Small Arms Survey Special Report, July 2001, http://www.smallarmssurvey.org/files/sas/publications/spe_reports_pdf/2001-sr1-humanitarian.pdf.

Muggah, Robert and Keith Krause (2009) 'Closing the Gap between Peace Operations and Post-Conflict Insecurity: Towards a violence reduction agenda', *International Peacekeeping*, 16(1): 136-50.

Mungai, Roselyn (2006) 'Demand Reduction in Action', http://www.iansa.org/un/review2006/presentations/Oxfam-GB-Kenya-Demand.pdf (27 July 2006).

Muppidi, Himadeep (1999) 'Postcoloniality and the Production of International Insecurity: The persistent puzzle of US–Indian relations', in Jutta Weldes, Mark Laffey, Hugh Gusterson and Raymond Duvall (eds.) *Cultures of Insecurity. States, Communities, and the Production of Danger* (Minneapolis: University of Minnesota Press), pp. 119-46.

Muppidi, Himadeep (2004) *The Politics of the Global* (Minneapolis: University of Minnesota Press).

Mutimer, David (2000) *The Weapons State. Proliferation and the Framing of Security* (Boulder, CO: Lynne Rienner).

Neuman, Stephanie G. (1984) 'International Stratification and Third World Military Industries', *International Organisation*, 38(1): 167-97.

Nonneman, Gerd (2001) 'Saudi–European Relations 1902-2001: A pragmatic quest for relative autonomy', *International Affairs*, 77(3): 631-61.

O'Connell, Dominic (2005) 'What Price Defence?' *Management Today*, 3 October 2005.

O'Grady, Margaret E. (1999) 'Small Arms and Africa', September 1999, http://www.caat.org.uk/resources/publications/countries/africa-0999.php (27 September 2009).

Ó Tuathail Gearóid (1996) *Critical Geopolitics. The Politics of Writing Global Space* (Minneapolis: University of Minnesota Press).

Øberg, Jan (1980) 'The New International Military Order: A Threat to Human Security', in Asbjørn Eide and Marek Thee (eds.) *Problems of Contemporary Militarism* (London: Croom Helm), pp. 47-74.

Omitoogun, Wuyi (2003) 'The Processes of Budgeting for the Military Sector in Africa', in SIPRI (ed.) *SIPRI Yearbook 2003. Armaments, Disarmament and Security* (Oxford: Oxford University Press), pp. 261-78.

Oxfam (1997) 'Oxfam Launches Campaign to Stop Small Arms Falling into Small Hands', Oxfam GB News Release, 13 December 1997, http://www.oxfam.org.uk/whatnew/press/confl.htm (22 November 2003).

Oxfam (1998) 'Small Arms, Wrong Hands', Oxfam UK Policy Paper, April 1998, http://www.oxfam.org.uk/policy/papers/smarms/exec.htm (29 October 2003).

Oxfam (2000) 'Oxfam Welcomes Decision to Stop Underwriting Arms Sales to World's Poor', 11 January 2000, http://www.oxfam.org.uk/whatnew/press/gordonbrown2.htm (29 October 2003).

Oxfam (2001), 'Up in Arms: Controlling the international trade in small arms', Oxfam GB Paper for the UN Conference on the Illicit Trade in Small Arms and Light Weapons in All Its Aspects July 2001, http://www.oxfam.org.uk/what_we_do/issues/conflict_disasters/downloads/upinarms.rtf (3 July 2008).

Oxfam (2002a) 'Words to Deeds. A New International Agenda for Peace and Security: Oxfam's 10-Point Plan', Oxfam Briefing Paper no. 14, August 2002, http://www.oxfam.org.uk/resources/policy/conflict_disasters/downloads/bp14_peace.pdf (3 July 2008).

Oxfam (2002b) 'The Spoils of Peace. How can tighter arms export controls benefit

both the poor and British industry', Briefing Paper, February 2002, http://www.
oxfam.org.uk/resources/policy/conflict_disasters/downloads/bp13_peace.rtf (3
July 2008).

Oxfam (2003) 'Oxfam GB-funded Peacebuilding Initiatives in the Arid Districts
of Kenya: Lessons and Challenges', March 2003, http://www.oxfam.org.uk/
what_we_do/issues/pastoralism/downloads/peacebuildingkenyafinal2004.pdf
(28 July 2006).

Oxfam (2006) 'Kenya: Programme overview', January 2006, http://www.oxfam.
org.uk/what_we_do/where_we_work/kenya/programme.htm (27 July 2006).

Oxfam (2008a) 'Shooting Down the MDGs. How irresponsible arms transfers
undermine development goals', Oxfam Briefing Paper, no. 120, October 2008.

Oxfam (2008b) 'Arms Transfer Decisions: Considering development', Oxfam
Briefing Note, June 2008.

Oxfam (2009) 'Practical Guide: Applying sustainable development to arms transfer
decisions', Oxfam Technical Brief, April 2009.

Oxfam (no date) 'Oxfam: Where the money comes from and where it goes', http://
www.oxfam.org.uk/about_us/downloads/moneytalk0304.pdf (1 July 2005).

Panitch, Leo and Sam Gindin (2003) 'Global Capitalism and American Empire',
in Leo Panitch and Colin Leys (eds.) *Socialist Register 2004. The New Imperial
Challenge* (London: Merlin Press), pp. 1-42.

Parekh, Bikhu (1995) 'Liberalism and Colonialism: A critique of Locke and Mill',
in Jan Nederveen Pieterse and Bhikhu Parekh (eds.) *The Decolonization of
Imagination. Culture, Knowledge and Power* (London: Zed Books), pp. 81-98.

Parmar, Inderjeet (2002) 'American Foundations and the Development of
International Knowledge Networks', *Global Networks* 2(1): 13–30.

Pasha, Mustapha Kamal and David L. Blaney (1998) 'Elusive Paradise: The promise
and peril of global civil society', *Alternatives*, 23(4): 417-50.

Percival, Jenny (2006) 'Q&A on the Saudi Arms Inquiry', *The Times Online*, 15
December 2006, http://www.timesonline.co.uk/tol/news/world/middle_east/
article755424.ece (3 October 2009).

Perlo-Freeman, Sam (2009a) 'Arms Production', in SIPRI (eds.) *SIPRI Yearbook
2009. Armaments, Disarmament and International Security* (Oxford: Oxford University Press), pp. 259-84.

Perlo-Freeman, Sam (2009b) 'Arms Transfers to the Middle East', SIPRI Background Paper, July 2009, http://books.sipri.org/files/misc/SIPRIBP0907.pdf
(10 December 2009).

Perlo-Freeman, Sam, Catalina Perdomo, Elisabeth Sköns and Petter Stålenheim
(2009) 'Military Expenditure', in SIPRI (eds.) *SIPRI Yearbook 2009. Armaments,
Disarmament and International Security* (Oxford: Oxford University Press), pp.
179-212.

Perlstein, Gary R. (2003) 'The World is not a Peaceful Place', *SAIS Review*, 23(1):
313-15.

Peters, Rebecca and Wayne LaPierre (2004) 'IANSA v The NRA', transcript of
debate between Peters and LaPierre, London, October 2004, http://www.iansa.
org/actin/nra_debate.htm (13 August 2009).

Petras, James (1999) 'NGOs: In the service of imperialism', *Journal of Contemporary
Asia*, 29(4): 429-40.

Phythian, Mark (2000a) *The Politics of British Arms Sales since 1964* (Manchester:

Manchester University Press).

Phythian, Mark (2000b) 'The Illicit Arms Trade: Cold War and Post-Cold War', *Crime, Law and Social Change*, 33, 1-52.

Pinter, Frances (2003) 'Funding Global Civil Society Organisations', in Mary Kaldor, Helmut Anheier and Marlies Glasius (eds.) *Global Civil Society 2003* (London: Sage), pp. 419-21.

Polden-Puckham Charitable Foundation (no date) website, http://www.polden-puckham.org.uk/ (3 July 2008).

Powell, Colin (2001) 'Remarks to the Foreign Policy Conference for Leaders of Nongovernmental Organisations', 26 October 2001, http://avalon.law.yale.edu/sept11/powell_brief31.asp (4 August 2009).

Press Association (2010a) 'BAE Systems pays £286m over corruption charges', *The Independent*, 5 February 2010.

Press Association (2010b) 'Lord Goldsmith defends deal for BAE', *The Guardian*, 7 February 2010.

Price, Richard (1998) 'Reversing the Gun Sights: Transnational civil society targets land mines', *International Organization*, 52(3): 613-44.

Price, Richard (2003) 'Transnational Civil Society and Advocacy in World Politics', *World Politics*, 55, pp. 579-606.

Prichard, Ian (2008) 'BAE's Frantic Flag-waving', *CAAT News*, April–May 2008, http://www.caat.org.uk/caatnews/Issues/2008_04/207BAELatest2.php (20 May 2008).

Reisinger, Sue (2008) 'In BAE Probe, US Steps in Where Brits Fear to Tread', 20 November 2008, *Law.Com*, http://www.law.com/jsp/ihc/PubArticleIHC.jsp?id=1202426158402 (5 August 2009).

Richards, Paul (1996) *Fighting for the Rainforest: War, youth and resources in Sierra Leone* (Oxford: International African Institute, in association with James Currey).

Robinson, William (1996) *Promoting Polyarchy. Globalization, US Intervention, and Hegemony* (Cambridge: Cambridge University Press).

Robinson, William I. (1999) 'Latin American in the Age of Inequality: Confronting the New "Utopia"', *International Studies Review*, 1(3): 41-67.

Robinson, William I. (2004) *A Theory of Global Capitalism. Production, Class, and State in a Transnational World* (London: Johns Hopkins University Press).

Roeber, Joe (2005) *Parallel Markets. Corruption in the International Arms Trade*, Goodwin Paper no. 3, June 2005 (London: CAAT).

Rosen, Steven (ed.) (1973) *Testing the Theory of the Military-Industrial Complex* (Lexington: Lexington Books).

Ross, Andrew (1989) 'Full Circle: Conventional proliferation, the international arms trade and Third World exports', in Kwang-il Baek, Ronald D. McLaurin and Chung-in Moon (eds.) *The Dilemma of Third World Defense Industries* (Boulder: Westview Press), pp. 1-31.

Rupert, Mark (1998) '(Re-)Engaging Gramsci: A response to Germain and Kenny', *Review of International Studies*, 24(3): 427-34.

Rupert, Mark (2005) 'Class Powers and the Politics of Global Governance', in Michael Barnett and Raymond Duvall (eds.) *Power in Global Governance* (Cambridge: Cambridge University Press), pp. 205-28.

Saferworld (2000) 'Memorandum from Saferworld (1 December 2000)', in Defence Committee, *First Report*, 14 February 2001, HC 115 (London: The Stationery

Office).

Saferworld (2002a) *Independent Audit of the 2000 UK Government Annual Report on Strategic Export Controls* (London: Saferworld).

Saferworld (2002b) 'Tighter Controls on Arms Exports Needed', 25 November 2002, http://www.saferworld.org.uk/PR1102.htm (21 November 2003).

Saferworld (2002c) 'The Missing Link? Arms exports and Labour's foreign policy', May 2002, http://www.saferworld.org.uk/briefLabour.htm (21 November 2003).

Saferworld (2003a) *Independent Audit of the 2001 UK Government Annual Report on Strategic Export Controls* (London: Saferworld).

Saferworld (2003b) 'Government Human Rights Report – Arms sales undermining human rights policy', 18 September 2003, http://www.saferworld.org.uk/PRhumanrights03.htm (20 May 2004).

Saferworld (2004a) 'House of Commons Debate on "Defence Procurement"', briefing for parliamentarians, 4 November 2004, http://www.saferworld.org.uk/publications.php/165/defence_procurement (19 May 2008).

Saferworld (2004b) *An Independent Audit of the 2002 UK Government Annual Report on Strategic Export Controls* (London: Saferworld).

Saferworld (2005) *An Independent Audit of the UK Government Reports on Strategic Export Controls for 2003 and the first half of 2004* (London: Saferworld).

Saferworld (2007a) *The Good, the Bad and the Ugly. A decade of Labour's arms exports*, May 2007 (London: Saferworld).

Saferworld (2007b) 'Memorandum from Saferworld', 3 July 2008, in Business and Enterprise, Defence, Foreign Affairs and International Development Committees, *Scrutiny of Arms Export Control (2008): UK Strategic Export Controls Annual Report 2006, Quarterly Reports for 2007, licensing policy and review of export control legislation*, HC254 (London: The Stationery Office), Ev. 73-78.

Saferworld (2007c) 'Saferworld Welcomes Decision to Close DESO', 26 July 2007, http://www.saferworld.org.uk/newslist.php/365/saferworld_welcomes_decision_to_close_deso?action=article&id=365 (8 August 2009).

Saferworld (2009a) *The Arms Trade Treaty. Countering Myths and Misperceptions*, July 2009 (London: Saferworld).

Saferworld (2009b) 'Promoting Chinese Support for Peace and Security in Africa', 10 September 2009, http://www.saferworld.org.uk/newslist.php/469/promoting_chinese_support_for_peace_and_security_in_africa?action=article&id=469 (25 November 2009).

Saferworld (no date, a) 'Arms Transfer Controls', http://www.saferworld.org.uk/pages/arma_transfer_controls_page.html (3 July 2008).

Saferworld (no date, b) 'UK Arms Transfer Controls', http://www.saferworld.org.uk/pages/uk_arms_transfer_controls.html (3 July 2008).

Saferworld (no date, c) 'About Us', http://www.saferworld.org.uk/pages/about_us.html (3 July 2008).

Saferworld (no date, d) 'Small Arms and Light Weapons', http://www.saferworld.org.uk/en/small_arms_intro.html (3 July 2008).

Saferworld (no date, e) 'National Small Arms Control', http://www.saferworld.org.uk/en/nat_sa_policy.html (3 July 2008).

Saferworld (no date, f) 'Small Arms Mappings', http://www.saferworld.org.uk/en/

mappings.html (3 July 2008).

Saferworld (no date, g) 'Action Against Small Arms', http://www.saferworld.org.uk/publications.php?id=59 (3 July 2008).

Said, Edward (1978) *Orientalism* (New York: Random House).

Sarkesian, Sam (ed.) (1972) *The Military-Industrial Complex. A Reassessment* (Beverly Hills: Sage Publications).

Saull, Richard (2005) 'Locating the Global South in the Theorisation of the Cold War: Capitalist development, social revolution and geopolitical conflict', *Third World Quarterly*, 26(2): 253-81.

SBAC (no date) 'National Defence Industries Council', http://www.sbac.co.uk/pages/82226198.asp#aGroup_1 (16 April 2009).

Schofield, Steven (2006) *The UK Defence Industrial Strategy and Alternative Approaches*, BASIC Occasional Papers on International Security Policy no. 50, March 2006, http://www.basicint.org/pubs/Papers/BP50.htm (3 July 2008).

Schofield, Steven (2007) *Oceans of Work: Arms conversion revisited*, 24 January 2007, BASIC Report, http://www.basicint.org/nuclear/beyondtrident/oceans.pdf (21 May 2008).

Schofield, Steven (2008) *Making Arms, Wasting Skills. Alternatives to militarism and arms production*, April 2008 (London: CAAT).

Scholte, Jan Aart (2002) 'Civil Society and Democracy in Global Governance', *Global Governance*, 8: 281-304.

Scholte, Jan Aart (2004) 'Civil Society and Democratically Accountable Global Governance', *Government and Opposition*, 39(2): 211-33.

Secretaries of State for Defence, Foreign and Commonwealth Affairs, International Development and Trade and Industry (2004) *Strategic Export Controls: Annual Report for 2002, Licensing Policy and Parliamentary Scrutiny. Response of the Secretaries of State for Defence, Foreign and Commonwealth Affairs, International Development and Trade and Industry*, October 2004 (London: The Stationery Office).

Secretary of State for Foreign and Commonwealth Affairs (2001) *Framework Agreement between the French Republic, the Federal Republic of Germany, the Italian Republic, the Kingdom of Spain, the Kingdom of Sweden, and the United Kingdom of Great Britain and Northern Ireland, concerning Measures to Facilitate the Restructuring and Operation of the European Defence Industry*, June 2001, Cm5185 (London: The Stationery Office).

Security Dialogue (2004) 'Special Section: What is "Human Security"?' *Security Dialogue*, 35(3): 345-87.

Shaw, Martin (1984a) 'Introduction: War and social theory', in Shaw, Martin (ed.) (1984) *War, State and Society* (London: Macmillan), pp. 1-24.

Shaw, Martin (1988) *Dialectics of War. An Essay in the Social Theory of Total War and Peace* (London: Pluto Press).

Shaw, Martin (1991) *Post-Military Society. Militarism, Demilitarization and War at the End of the Twentieth Century* (Philadelphia: Temple University Press).

Shaw, Martin (1994) 'Civil Society and Global Politics. Beyond a Social Movements Approach', *Millennium*, 23(3): 647-67.

Shaw, Martin (1997) 'The State of Globalization: Towards a theory of state transformation', *Review of International Political Economy*, 4(3): 497-513.

Shaw, Martin (2002) 'Post-Imperial and Quasi-Imperial: State and empire in the global era', *Millennium: Journal of International Studies*, 31(2): 327-36.

Shaw, Martin (2005) *The New Western Way of War. Risk-transfer War and Its Crisis in Iraq* (Cambridge: Polity).

Short, Clare (2003) 'Small Arms and Light Weapons', speech delivered at Lancaster House, 14 January 2003, http://62.189.42.51.DFIDstage/news/Speeches/files/sp14jan03.html (28 July 2004).

Singh, Jasjit (ed.) (1995) *Light Weapons and International Security* (London and New Delhi: BASIC and Indian Pugwash Society).

SIPRI (2008) 'Recent Trends in Military Expenditure', http://www.sipri.org/contents/milap/milex/mex_trends.html (30 April 2009).

Sklair, Leslie (1997) 'Social Movements for Global Capitalism: The Transnational Capitalist Class in Action', *Review of International Political Economy*, 4(3): 514-38.

Small Arms Survey (2001) *Small Arms Survey 2001. Profiling the Problem* (Oxford: Oxford University Press).

Small Arms Survey (2003) *Small Arms Survey 2003. Development Denied* (Oxford: Oxford University Press).

Small Arms Survey (2004) *Small Arms Survey 2004. Rights at Risk* (Oxford: Oxford University Press).

Small Arms Survey (2007) 'The Militarization of Sudan. A preliminary review of arms flows and Holdings', Sudan Issue Brief, Human Security Baseline Assessment, Number 6, April 2007, http://www.smallarmssurvey.org/files/portal/spotlight/sudan/Sudan_pdf/SIB%206%20militarization.pdf (19 November 2009).

Small Arms Survey (2009) *Small Arms Survey 2009. Shadows of War* (Cambridge: Cambridge University Press).

Sörensen, Jens Stilhoff (2002) 'Balkanism and the New Radical Interventionism: A structural critique', *International Peacekeeping*, 9(1): 1-22.

Spear, Joanna and Neil Cooper (2006) 'The Defence Trade', in Alan Collins (ed.) *Contemporary Security Studies* (Oxford: Oxford University Press), pp. 311-29.

Stavrianakis, Anna (2005) '(Big) Business as Usual. Sustainable Development, NGOs and UK Arms Export Policy', *Conflict, Security and Development,* 5(1), April 2005, 45-67.

Stavrianakis, Anna (2006) 'Call to Arms: The university as a site of militarised capitalism and a site of struggle', *Millennium*, 35(1): 139-54.

Stavrianakis, Anna (2008) 'Licensed to Kill: the United Kingdom's arms export licensing process', *The Economics of Peace and Security Journal*, 3(1), 32-9.

Stearman, Kaye (2009) 'Weapons of Economic Destruction', *The Guardian,* 8 September 2009, http://www.guardian.co.uk/commentisfree/2009/sep/08/excel-arms-fair-defence-spending (10 September 2009).

Stohl, Rachel (2001) 'United States Weakens Outcome of UN Small Arms and Light Weapons Conference', *Arms Control Today*, September 2001, pp. 34-5.

Straw, Jack (2002) *Hansard*, Written Answers to Questions, 8 July 2002, col. 652-4W. http://www.publications.parliament.uk/pa/cm200102/cmhansrd/vo020708/text/20708w01.htm#20708w01.html_spnew3 (22 June 2009).

Sullivan, Sian (2005) '"We are heartbroken and furious!" Violence and the (anti)-globalisation movement(s)', in Catherine Eschle and Bice Maiguaschca (eds.) *Critical Theories, International Relations and 'the Anti-Globalisation Movement'. The Politics of Global Resistance* (Abingdon: Routledge) pp. 174-94.

Tapol (2003) 'Call for an International Military Embargo against Indonesia', 23 June

2003, http://tapol.gn.apc.org/news/files/st030623.htm (16 May 2006).

ThisDay (2007) 'Overpriced Helicopters: Radar men also linked to 25bn/-choppers deal for the military', *ThisDay*, 3 May 2007, http://www.thisday.co.tz/News/1886.html (14 August 2009).

Thomas, Gareth, MP (2006) 'UK Statement at the 2006 Review Conference of the UN Programme of Action to Prevent, Combat and Eradicate the Illicit Trade in Small Arms and Light Weapons in All Its Aspects', 26 June 2006, http://www.un.org/events/smallarms2006/pdf/arms060626uk-eng.pdf (25 February 2008).

Tilly, Charles (1985) 'War Making and State Making as Organized Crime', in Peter Evans, Dietrich Rueschmeyer and Theda Skocpol (eds.) *Bringing the State Back In* (Cambridge: Cambridge University Press), pp. 169-91.

Tomlinson, Heather (2003) 'MoD Axe Could Fall on 10,000 Jobs', *The Independent*, 15 June 2003.

Turner, Mike (2007) Letter to Gordon Brown, 26 July 2007, http://www.defenseindustrydaily.com/files/DESO_Letter_BAE_2007-07-26.pdf (28 November 2008).

UK Mission to the United Nations (no date) 'UK Implementation of and Support for the UN Programme of Action on SALW', http://www.ukun.org/UNPoA.pdf (28 July 2004).

UKWG (2006) 'Memorandum to the Quadripartite Committee: The future of the Export Control Organisation', http://www.saferworld.org.uk/publications.php/158/ukwg_on_arms_submission_on_export_control_organisation_eco (Janauary, accessed 19 May 2008).

Unite (no date) *Maintaining a World Class UK Defence Industry*, http://www.unitetheunion.com/pdf/Maintaining%20World%20Class%20UK%20Defence%20Industry%20(JN504)%20Final.pdf (10 April 2009).

United Nations (2001) *Programme of Action to Prevent, Combat and Eradicate the Illicit Trade in Small Arms and Light Weapons in All Its Aspects*, http://disarmament.un.org/cab/poa.html (2 August 2006).

United Nations General Assembly (2006) *Towards an Arms Trade Treaty: Establishing common international standards for the import, export and transfer of conventional arms*, Resolution 61/89, 18 December 2006 (New York: United Nations General Assembly).

US Department of State (2007) 'The US–UK Defence Trade Cooperation Treaty', Fact Sheet, Bureau of Political-Military Affairs, Washington, DC, 10 August 2007, http://www.state.gov/t/pm/rls/fs/90740.htm (18 November 2008).

Vogel, Ann (2006) 'Who's Making Global Civil Society: Philanthropy and the US empire in world society', *The British Journal of Sociology*, 57(4): 635-55.

Wall, John and Rob Johnston (2004) *Maintaining A Critical Mass for UK Defence*, http://www.amicustheunion.org/pdf/Amicus_Defence%20final.pdf (16 May 2005).

Wallace, Tina (2003) 'NGO Dilemmas: Trojan horses for global neoliberalism?' in Leo Panitch and Colin Leys (eds.) *Socialist Register 2004. The New Imperial Challenge* (London: Merlin Press), pp. 202-19.

Waqo, Halakhe D. (2003) 'Peacebuilding and Small Arms: Experiences from Northern Kenya', presentation to workshop at UN Biennial Conference of States on Small Arms Programmes of Action, New York, www.iansa.org/un/notes/peacebuilding_and_small_arms.doc (17 March 2010).

Watkins, Kevin (2001) 'This Deal Is Immoral, Mr Blair', *The Guardian*, 21 December 2001.

Weaver, Matthew (2009) 'G20 Protestors Blasted by Sonic Cannon', *The Guardian*, 25 September 2009.

Wendt, Alexander and Michael Barnett (1993) 'Dependent State Formation and Third World Militarization', *Review of International Studies*, 19(4): 321–47.

van der Westhuizen, Janis (2005) 'Arms over AIDS in South Africa: Why the boys had to have their toys', *Alternatives*, 30: 275–95.

Weber, Max (1991) 'Science as a Vocation', in H.H. Gerth and C. Wright Mills (eds.) *From Max Weber: Essays in Sociology* (London: Routledge), pp. 129–56.

Wezeman, Siemon T., Mark Bromley and Pieter D. Wezeman (2009) 'International Arms Transfers', in SIPRI (eds.) *SIPRI Yearbook 2009. Armaments, Disarmament and International Security* (Oxford: Oxford University Press), pp. 299–320.

Willett, Susan (1999) 'The Arms Trade, Debt and Development', http://www.caat.org.uk/information/publications/economics/debt-and-development-0599.php (21 November 2003).

Williams, Jody (1999) 'The International Campaign to Ban Landmines – A model for disarmament initiatives?' 3 September 1999, http://nobelprize.org/nobel_prizes/peace/articles/williams/index.html (15 November 2009).

Williams, Tim (2007) 'Exporting Arms, Importing Standards', *Defence Management Journal*, August 2007, accessed via http://www.sbac.co.uk/pages/59626406.asp (16 April 2009).

Wood, Ellen Meiksins (1990) 'The Uses and Abuses of "civil society"', in Ralph Miliband, Leo Panitch and John Saville (eds.) *Socialist Register 1990. The Retreat of the Intellectuals* (London: Merlin Press), pp. 60–84.

Zelter, Angie (2004) 'Civil Society and Global Responsibility: The arms trade and East Timor', *International Relations*, 18(1): 125–40.

Index

public opinion, 67, 69–70

Qatar, 120
Quaker Peace and Social Witness, 6

RAF, Hawk jets, 48–9, 101
Red Diamond, 121
reformists: can undermine transformist vision, 92; Control Arms, 56, 135; insider strategy, 63; persuasive rather than confrontational, 168; seek tighter regulation, 9, 33, 34–7; see also Amnesty International; Oxfam; Saferworld
Robinson, William I., 16, 20, 27, 45, 53–4, 127, 134, 167, 178–9, 203
Rockefeller Foundation, 73
Roeber, Joe, 120
Romania, 120
Rupert, Mark, 9, 26–7, 203
Russia, 40, 46, 78, 126, 140, 144
Rwanda, 138

Safer Albania, 174
Safer Rwanda, 174
SaferAfrica, 148, 174
Saferworld: arms trade employment, 95; Audits, 78, 100–101; background, 6; Biting the Bullet campaign, 147; capacity building, 149–50; DESO closure, 98–9; DfID, 85, 123; economics of arms trade, 94–5; European defence, 108–110; funding, 72–3, 74–5; incorporation issue, 106; Indonesia, 128, 130; insider strategy, 63, 64–5, 147–8; intra-Northern trade, 100–101, 107–8, 111–13; objectives, 34, 35–6; small arms, 138, 147–8; small arms exports, 140; UK arms exports, 116; UK defence procurement, 101–2; UK–US relations, 104–5
Saudi Arabia: arms trade SFO investigation, 1, 2, 91, 120–21, 125–6; military purchases, 55; small arms exports, 140
Save the Children, 76
Schofield, Steven, 103
Scholte, Jan Aart, 16, 24, 26
security: hard and soft, 14; international security, 49–51, 163–4; security sector reform (SSR), 142; and small arms, 52, 138, 139, 141–3, 155–6, 157–8; see also human security
Serious Fraud Office (SFO), 1, 2, 45, 86, 91, 120–21, 125–6
Shaw, Martin, 16, 43, 45, 48, 54–5, 107, 172, 177–9, 205

Short, Clare, 125
Singapore, 36
small arms: Biting the Bullet project, 65, 141, 145, 147; civilian ownership, 42, 154, 160; conflict prevention, 42, 54, 59–60, 136–7, 145–6, 158; deaths, 40, 153; definition, 61; and global civil society, 149–55; and human security, 155–6; illicit trade, 42, 143–4, 144–5, 149; international action, 143–4, 147–8, 159–60; legitimate weapons, 146; NGO influence, 137–43, 155–6, 160–62, 164; NGOs as nodes, 146–9; and poverty, 139; regional variations in interventions, 159–60; removal from non-state actors, 154, 160; security and development, 50, 156; social transformation and state-building, 157–9; state control of, 160–61; supply, 41–2, 49–50, 139–41, 143–5; UK government, 64–5, 146–7; and violence, 35, 49, 50–51, 141, 152, 155, 179–80; see also IANSA (International Action Network on Small Arms)
Small Arms Survey, 5, 41–2, 61, 139, 140, 143, 147, 152, 154, 156, 159
social transformation, 157–8
South Africa, 36, 120, 140, 151
South Korea, 36
Southeast Asia, 153
SPEAK campaign group, 7, 97
SSR (security sector reform), 142
state: and arms capital, 38, 39, 57, 59, 86–8, 165–6; and civil society, 15–19, 147–8
state formation, 29, 31, 55, 127, 156, 175, 177–8, 180, 187, 198
strategies: abolitionist, 55–6; complementarity of, 91–2; see also insider strategies; outsider strategies; reformists; transformists
Straw, Jack, 106, 107
Sudan, 69, 79, 131, 139–40, 149, 183, 206
sustainable development, 115–21, 122, 123, 126
Sweden, 140
Switzerland, 140
Syria, 126, 140, 144

Tajikistan, 132
Tanzania: air traffic control system, 89, 115–16, 118, 120–21, 125, 173; small arms, 155, 173
TAPOL campaign group, 7, 131
Tebbit, Sir Kevin, 86
terrorism, 143; see also War on Terror

www.ingramcontent.com/pod-product-compliance
Lightning Source LLC
Chambersburg PA
CBHW022312280326
41932CB00010B/1069

* 9 7 8 1 8 4 8 1 3 2 6 9 6 *